THE MYSTERY OF THE
TUNGUSKA
FIREBALL

SURENDRA VERMA

ICON BOOKS

Originally published in 2005 by Icon Books Ltd.

This edition published in the UK in 2006 by Icon Books Ltd,
The Old Dairy, Brook Road, Thriplow, Cambridge SG8 7RG
e-mail: info@iconbooks.co.uk
www.iconbooks.co.uk

Sold in the UK, Europe, South Africa and Asia
by Faber and Faber Ltd, 3 Queen Square,
London WC1N 3AU
or their agents

Distributed in the UK, Europe, South Africa and Asia
by TBS Ltd, Frating Distribution Centre, Colchester Road,
Frating Green, Colchester CO7 7DW

Published in Australia in 2006
by Allen & Unwin Pty Ltd,
PO Box 8500, 83 Alexander Street,
Crows Nest, NSW 2065

Distributed in Canada by
Penguin Books Canada,
90 Eglinton Avenue East, Suite 700,
Toronto, Ontario M4P 2YE

ISBN-10: 1-84046-728-2
ISBN-13: 978-1840467-28-4

Typesetting by Hands Fotoset

Printed and bound in the UK by
Bookmarque Ltd.

CONTENTS

Surendra Verma is a science writer based in Melbourne, Australia since 1970. He has published several books and numerous articles, besides churning out pretentious prose for corporate and government publications. His *Little Book of Scientific Principles, Theories and Things* was published in 2005 by New Holland.

ACKNOWLEDGEMENTS

During the writing of this book, I requested the major proponents of two leading theories on the Tunguska event, Dr Zdenek Sekanina of NASA's Jet Propulsion Laboratory in California and Dr Vitalii Bronshten of the Committee of Meteorites of the Russian Academy of Sciences, to express their latest views on their theories. I was saddened to hear of Dr Bronshten's death a few weeks after I had written to him. I'm grateful to Dr Sekanina for his response.

I'm also grateful to Dr Robert Foot of the University of Melbourne; Professor Wolfgang Kundt of the University of Bonn; and Dr Kiril Chukanov of Chukanov Quantum Energy in Salt Lake City, Utah, for commenting on their research on the Tunguska event.

My special thanks go to Dr Marek Zbik of the Ian Wark Research Institute at the University of South Australia for his help in providing research papers and illustrations; to Mr Vitalii Romeiko of Moscow for giving permission to use his photograph of the Suslov crater; to the staff of the State Library of Victoria for their courteous help on numerous occasions; to Geoff

Coleman, Simon Kwok, Ruth Learner, Darren Lewin-Hill and Arun Tomar for their moral support; and to Eric ('Fizzle') Fiesley and Colin ('Stick') Storer for being true Aussie mates for more than 30 years.

The writing of this book became a pleasure with Dr Andrei Ol'khovatov's enthusiastic support in many ways. Dr Ol'khovatov, formerly of the Soviet Radio Instrument Industry Research Institute and now an independent researcher in Moscow, is a well-known personality in the Tunguska cyberspace and research communities. Thank you, Andrei. Keep the Tunguska fireball burning brightly.

I'm indebted to Icon Books' publisher Simon Flynn for giving me the opportunity to write this book, and their editor Duncan Heath for his indispensable advice.

Finally, I would like to thank my wife Suman and my sons Rohit and Anuraag for their unfailing support throughout the writing of this book.

INTRODUCTION

The so-called Tunguska event has been part of the folk-lore of science since 1927, when Leonid Kulik became the first scientist to visit the explosion site. He saw an oval plateau 70 kilometres wide where the forest had been flattened. Trees were not uprooted: instead they were stripped of their branches, snapped off and scattered like matchsticks pointing away from the direction of the blast. Even after a careful search Kulik found no crater or other evidence of impact. He searched for meteorite fragments but found nothing. As there was no impact crater and no substantial remnants, a giant meteorite could not have caused the Tunguska explosion. If it were not a meteorite, then what caused the explosion?

The search for the answer to this question has generated a Tunguska industry that has kept both scientists and charlatans busy for eight decades.

Hundreds of research articles in renowned and not-so-renowned journals prove that bright scientific minds are keen to solve the riddle. Tunguska also provides them with an opportunity to test-drive new theories: black holes, ball lightning, anti-matter and mirror matter are some of the examples. Astronomers' recent fascination with the probability of a rogue asteroid striking Earth has also refocused scientific attention on Tunguska. Adventurous ones can always make a trip to Tunguska and look for the evidence for 'stones from the heavens'. There are workshops, symposia and conferences for the non-adventurous ones.

Numerous websites, conspiracy theories, sensational TV documentaries, one episode of the popular TV series *The X-Files*, and so on prove that Tunguska adds enough exotic spices to make science palatable. For science fiction fans, Tunguska has boundless skies for spaceships to roam.

Whichever way you look, Tunguska is a fascinating journey in serious science, speculative science and science fiction. *The Tunguska Fireball* attempts to give you a glimpse of that journey.

A note on units and terms

Metric units are used in this book. Even if you are not familiar with them, this should not diminish your enjoyment of reading or your understanding of the subject matter. You may assume metres to be yards, kilometres miles, Celsius Fahrenheit and tonnes tons, although values used are accurate. The energy unit 'megaton' is explained on its first appearance in the text.

The term 'theory' is used in a general sense and includes terms such as 'hypothesis' and 'model'.

CHAPTER ONE

FIRE IN THE SKY

About 7.14 a.m., 30 June 1908. The Central Siberian Plateau near the Stony Tunguska River, a remote and empty wilderness of swamps, bogs and hilly pine and cedar forests. Not a soul in sight for scores of kilometres. The eerie silence is punctuated by the shuffle of the hoofs of reindeer grazing in the morning sun and the hum of dense swarms of ferocious mosquitoes appropriately called 'flying alligators'.

Suddenly a blindingly bright pillar of fire, the size of a tall office building, races across the clear blue sky. The dazzling fireball moves within a few seconds from the south-southeast to the north-northwest, leaving a thick trail of light some 800 kilometres long. It descends slowly for a few minutes and then explodes about 8 kilometres above the ground. The explosion lasts only a few seconds but it is so powerful that it can be compared only with an atomic bomb – 1,000 Hiroshima atomic bombs.

The explosion flattens 2,150 square kilometres of the mighty taiga, stripping millions of ancient trees of leaves and branches, leaving them bare like telegraph poles and scattering them like matchsticks. A dark mushroom cloud of dust rises to a height of 80 kilometres over the area after the explosion. A black rain of debris and dirt

1

follows. Shortly afterwards, bluish clouds of ice-coated dust grains are seen against the red sky.

At Vanavara, a trading station about 70 kilometres from the explosion site, a trader, S.B. Semenov, sitting outside his house is knocked off his chair by violent shock waves. The explosion emits so much heat that it seems to be burning his shirt. He said later that he had only a moment to note the size of the bright blue 'tube' that covered an enormous part of the sky. 'Afterwards it became dark and at the same time I felt an explosion that threw me several feet from the porch and for a moment I lost consciousness.' He regains consciousness to hear a tremendous sound that shakes the whole house and nearly moves it off its foundation, breaks the glass in the windows and damages his barn considerably. The earth trembles and then the sky splits apart and a hot wind, as from a cannon, blows past the houses.

Another trader, P.P. Kosolopov, who is walking outside his house, feels his ears burning. He covers them with his hands and runs into his house. Inside the house, earth starts falling from the ceiling and the door of his large stove blows out. The window panes break and he hears thunder disappearing to the north. When it is quieter he goes into the yard but sees nothing else.

Several kilometres north of Vanavara, the tents of dozens of nomads and herdsmen, including the occupants, are blown up into the air by the shock waves that follow the explosion. They all suffer slight bruises when they fall back to the ground. An elderly man hits a tree and breaks his arm. Another elderly man dies of fright. Hundreds of reindeer belonging to four separate herds are killed as the pines and cedars around them blaze. Dense smoke envelops the forest.

Fishermen repairing their rafts along the banks of the Stony Tunguska River are thrown into the air. Their horses stumble and fall to the ground as the shock waves after the explosion pulse through the rocks.

A farmer ploughing his hillside land about 200 kilometres south of the explosion site hears sudden bangs, as if from gunfire. His horse falls on its knees. The fir forest around him is bent over by the wind. He seizes hold of his plough with both hands so that it is not carried away. The wind is so strong that it blows away most of the soil from the surface of the ground, and then drives a wall of water up the nearby river. A flame shoots up above the forest in the north.

About 600 kilometres to the southwest, the Trans-Siberian Express jars and shakes wildly on its tracks, built only three years earlier. Passengers are frightened by the loud bursts of noise. The startled driver sees the tracks ahead rippling. He brings the train to a screeching halt. Sounds of distant thunder follow.

Villagers in Znamenskoye, 700 kilometres from the explosion site, see bright lights in the sky. After the passage of the fireball, villagers in Achayevskoye, 1,200 kilometres from the site, hear loud explosions, like gunfire, which continue for several minutes.

Reverberations around the globe

The explosion was registered by an earthquake-measuring station some 4,000 kilometres away in the city of St

Petersburg. Atmospheric disturbances were also recorded by more distant stations around the world.

Disturbances in Earth's magnetic field – similar to ones produced by nuclear explosions in the atmosphere – were recorded 970 kilometres south of the explosion site by the Irkutsk Magnetic and Meteorological Observatory. The magnetic storm lasted more than four hours. Subsequent analysis of these records showed that the epicentre of the 'earthquake' coincided with the location of the explosion (latitude 60 degrees 55 minutes north, longitude 101 degrees 57 minutes east) and confirmed the accurate time of the event (0014 GMT; 7.14 a.m. local time).

Two weeks after the explosion, the Smithsonian Astrophysical Observatory and Mount Wilson Observatory in the United States recorded a marked decrease in the air's transparency. It has been suggested that this was due to the loss of vast amounts of material from the fireball as it passed through the atmosphere. Calculations showed this loss to be several million tonnes, a hundred times more than the normal yearly fall of meteorite matter on Earth.

On the evening of the explosion, bright, colourful and prolonged dusks were noticed across the Continent as far as Spain. Photographs of the sky taken on that night at the Heidelberg Astronomical Observatory were badly clouded because of the bright sky. A Hamburg photographer who took a picture of the sky at 11 p.m. described it as 'volcanic dust', as the memory of the 1883 Krakatoa volcano, modern history's most violent eruption, was still fresh in people's memory. After sunset in Antwerp the northern horizon appeared to be on fire. A report in *Vart Land*, a Stockholm evening newspaper, described a 'strange illumination' on the night of 30 June.

The nights were also unusually bright over the British Isles. On 2 July the London *Times* published a letter from a Miss Katharine Stephen of Huntingdon about 'the strange light in the sky' which she and her sister had observed between midnight and 12.15 a.m. on 1 July. 'It would be interesting', she requested, 'if anyone could explain the cause of so unusual a sight.' The next day a Holcombe Ingleby of Brancaster wrote about the 'curious sun effects at night' which had the appearance of a dying sunset of exquisite beauty. 'This not only lasted but actually grew both in extent and intensity till 2.30 this morning', he continued. 'I myself was aroused from sleep at 1.15, and so strong was the light at that hour that I could read a book by it in my chamber quite comfortably.' In the same issue the newspaper also reported that in Dublin 'a very remarkable afterglow prolonged the daylight to such an extent that it was possible to read a newspaper in the open air'.

On 4 July *The Times* made an attempt to explain 'the remarkable ruddy glows which have been seen on many nights lately'. The newspaper remarked that these glows had been seen over an area extending as far as Berlin, and pointed out that there was a considerable difference of opinion as to the nature of these glows: 'Some hold that they are auroral; their colour is quite consistent with this view ... [others hold that] the phenomenon was simply an abnormal twilight glow ... We may recall the circumstances of the wonderful glows which were seen in this country in the autumn of 1883, and which were due to the dust scattered in the upper atmosphere by the terrific outburst at Krakatoa at the end of August. Those glows had many points in common with the recent ones.' *The Times* noted that 'distance is no obstacle in vast cosmical phenomena of this kind, which are absolutely world-embracing' and it was possible that the dust may have come 'from some unreported volcanic eruption in some little-known region of the world'.

On 3 July, *The New York Times* reported that 'remarkable lights were observed in the northern heavens on Tuesday and Wednesday nights, the bright diffused white and yellow illumination continuing throughout the night until it disappeared at dawn', which the newspaper attributed to 'important changes on the sun's surface, causing electrical discharges'. Two days later the newspaper published another report from its London correspondent. The report, 'LIKE DAWN AT MIDNIGHT: LONDON SEES SKY BLUE AND CLOUDS TIPPED WITH PINK AT THAT HOUR', said that several nights through the week were marked by strange atmospheric effects. 'Following sunsets of exceptional beauty and twilight effects remarkable even in England', the report continued, 'the northern sky at midnight became light blue, as if the dawn were breaking, and the clouds were touched with pink, in so marked a fashion that police headquarters was rung up by several people, who believed a big fire was raging in the north of London.'

At Edinburgh Observatory the night sky was noted as being very striking, 'practically daylight'. This practical daylight, *The Scotsman* reported, caused shadows to be cast in rooms with windows facing north. The saying 'make hay while the sun shines' took on a new meaning when farmers in the north of England worked in the fields all night getting in their hay before an impending storm broke.

The night brightness slowly diminished and disappeared after a few days, but scientists continued to speculate on the cause of these 'nocturnal glows'. W.F. Denning, an astronomer from Bristol, wrote in the weekly journal *Nature* on 9 July: 'I have never seen June nights so dark, and Milky Way so gorgeously displayed in the heavens ... nor have I ever noticed the sky so bright as it appeared on the nights of June 30 and July 1.' Bohuslav Brauner of the Bohemian University in Prague also wrote in the same issue: 'The peculiar light phenomenon at midnight on June 30 ... was also observed by me at Prague ... It is reported that magnetic disturbances were experienced on the telegraphic lines, but I saw no trace of the characteristic auroral bands or columns.' The next week Denning wrote again, this time discounting his earlier explanation that the night glows were due to aurora borealis (displays of coloured lights in the northern skies): 'Whatever the true nature of the recent exhibition may have been, it is certain that something in the air exercised the capacity of reflection in a very high degree.'

In its August issue, the magazine of the Royal Observatory at Greenwich provided a vivid account of the night glows: 'At 9.30 p.m. at Greenwich, on June 30, the sky along the north-west and north horizon was of a brilliant red, in fact there was what is usually termed a "brilliant sunset", the only peculiarity being that the brightness stretched more to the north than is usual, and endured, so that at one o'clock in the morning it extended well across the north of the horizon, and the northern sky above was of a brightness approaching that of the southern sky at the time of Full Moon.' The unsigned article went on to say that observations failed to give any evidence that it was an auroral display but 'the light, indeed, was sufficient to take photographs of terrestrial objects'. An excellent photograph taken shortly before midnight with an exposure of about one minute showed the domes of the Naval College of the Royal Observatory with the College's training ship *Fame* in the foreground.

On the other side of the Atlantic, *Scientific American* reported on 29 August that 'sky glows', called by some of the European astronomers aurora displays, were now the subject of interesting discussion in astronomical circles, especially among the scientists of Europe. 'For some time a peculiar strong orange-yellow light over the horizon, the color of which was more orange in its lower parts and more yellow in its higher parts, has been observed all over northern Europe and the United States', the report continued. 'Clouds or spiral streams of various tints were brilliantly outlined across the sky, so luminous that few stars could be seen, and the Milky Way was hardly distinguishable.' The report quoted Denning and Brauner from *Nature* and added that 'both say they saw no trace of the characteristic auroral bands or columns in this phenomenon'.

The scientists at the Dublin meeting of the British

Association for the Advancement of Science in September 1908 did not know that the microbarograph invented in 1903 by two prominent members, W.N. Shaw (later Sir Napier Shaw) and W.H. Dines, had, in fact, recorded disturbances caused by the Tunguska explosion. The microbarograph automatically records sudden small changes in atmospheric pressure, but does not show changes due to ordinary rising and falling of the barometer. During a discussion on wave motion, as a curious example of atmospheric wave motion, Shaw presented six graphs recorded at six different locations in England on 30 June 1908 at about 0514 GMT (that is, five hours after the Tunguska explosion). Each graph showed a series of air waves during a period of about one hour. Each wave had four clear peaks, as if there had been four disturbances in Earth's atmosphere during that period. Shaw noted that the peaks lasted for about fifteen minutes and they were then 'violently interrupted by a sudden though slight disturbance' for a similar interval. Scientists at the meeting thought that this curious phenomenon was due to a large atmospheric disturbance in some unknown part of the world.

Like the night glows, the six graphs were to remain one of the unexplained mysteries of science for two decades. No one, except some observers in Siberia, was aware that a mysterious fireball had exploded in the Siberian sky.

The British Antarctic Expedition of 1907–09, led by Sir Ernest Shackleton, was wintering at the Cape Royds Station in the Antarctic when the Tunguska fireball exploded. Did Shackleton's party observe aurora australis

9

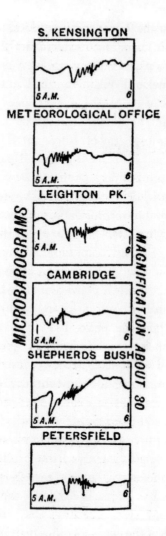

Figure 1: Six microbarographs recorded at different locations in England on 30 June 1908 at about 0514 GMT. (From F.J.W. Whipple, 'The great Siberian meteor', in the *Quarterly Journal of the Royal Meteorological Society*, vol. 56, 1930.)

(coloured lights in the far southern skies) at the time of the explosion? There is no evidence of this; however, there is a record of an exceptional aurora seven hours before the explosion. Was this aurora in any way related to the fireball?

The 'tongue of flame'

The Siberian newspapers of the time were no better informed than their distant British and American counterparts, but the explosion was reported widely.

In the Irkutsk newspaper *Sibir* on 15 July, a correspondent reported that on the morning of 30 June peasants in the village of Nizhne-Karelinskoye (about 465 kilometres from the explosion site) saw quite high above the horizon a body shining very brightly with a bluish-white light. The body was in the form of 'a pipe', and too bright for the naked eye. It moved vertically downwards for about ten minutes before it approached the ground and pulverised the forest. A huge cloud of black smoke was formed 'and a loud crash, not like thunder, but as if from the fall of large stones or from gunfire was heard. All the buildings shook and at the same time a forked tongue of flame broke through the cloud.' The incident frightened the villagers and they ran out into the street in panic. Everyone thought that the end of the world was approaching.

At the time of the explosion, the correspondent of the *Sibir* report was in Kirensk (about 500 kilometres away) and 'heard in the northwest what sounded like gunfire repeated at intervals at least ten times, and lasting in all about fifteen minutes'. He noted that the northwest-facing windows in several houses were shattered.

The newspaper *Golos Tomska* sent a correspondent to the town of Kansk (about 635 kilometres from the site), near which a large meteorite was rumoured to have fallen, to verify the information. On 17 July the newspaper said that 'all the details of the fall of the meteorite should be ascribed to the overactive imagination of impressionable people'. The newspaper, however, accepted that 'there is no doubt that a meteorite fell, probably some distance away, but its huge mass and so on are very doubtful'. On 28 July the newspaper reported an earthquake in Kansk on the morning of 30 June. The report said that the earthquake was followed by a subterranean crash and a roar as if from distant firing. 'Doors, windows and the lamps before icons were all shaken', the report continued. 'Five to seven minutes later a second crash followed, louder than the first, accompanied by a similar roar and followed after a brief interval by yet another crash, fainter than the first two.'

A detailed and dramatic description of the event appeared in the Krasnoyarsk newspaper *Krasnoyarets* on 26 July, from the paper's correspondent in the village of Kezhma (about 215 kilometres away). This is the only newspaper report to come from the inhabited place nearest to the explosion site. After a detailed description of 'a subterranean shock which caused buildings to tremble', the newspaper quoted witness reports that said that 'before the first bangs were heard a heavenly body of a fiery appearance cut across the sky from south to north ... neither its size nor shape could be made out owing to

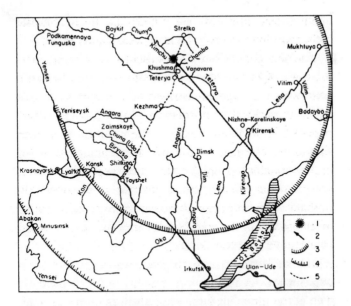

Figure 2: Map of Tunguska and the surrounding areas showing the extent of various eyewitness observations. Key: 1 – impact site; 2 – flight path of the fireball; 3 – extent of fireball visibility; 4 – extent of explosion sounds; 5 – track of the first Tunguska expedition. Approximate scale: 1 cm = 175 km. (Courtesy Marek Zbik, University of South Australia.)

its speed and particularly its unexpectedness'. The report went on to say that when the flying object touched the horizon a huge flame shot up that cut the sky in two. 'The bangs were heard as soon as the "tongue of flame" disappeared. On the island near the village the horses and cows became noisy and began running wildly about … Terrible shocks were heard coming from somewhere. They shook the earth, and their unknown source inspired a kind of superstitious terror. People were

13

literally dumbfounded.' The report also noted that the blaze must have lasted for at least a minute because it was noticed by many peasants in the fields.

The 'tongue of flame' or the 'tongue of fire' seems a common eyewitness description of the Tunguska fireball and it appears in many Siberian newspapers of the time. A report from the village of Nizhne-Ilimskoye (about 420 kilometres away) said that the population there and in the surrounding villages saw 'a fiery body like a beam' shoot from south to northwest before they heard the thunder. The fiery body disappeared immediately after the bang and a 'tongue of fire' appeared in its place, which was followed by smoke.

Many Russian newspapers outside Siberia also published news of the bright nights and shiny clouds seen at the time, but there was hardly any report on the fireball (there was no *Pravda* yet: it was started by Leon Trotsky in October 1908 in Vienna, moving to St Petersburg in 1912 and then to Moscow in 1918). However, on 4 July 1908 *The Trading Industrial Gazette* (St Petersburg) ran a short piece headlined: 'MORE ABOUT THE FALL OF THE METEORITE: YESTERDAY WE RECEIVED A TELEGRAM SAYING ONLY – THE NOISE WAS CONSIDERABLE BUT NO STONE FELL'.

The year 1908

In 1908 Russia was a country caught in political unrest and social upheaval. The investigation of an earthquake or a meteorite in remote Siberia was the last priority of the Russian authorities in the then capital St Petersburg.

The First World War and the Bolshevik Revolution were to occupy the country's attention for many years. The impact site was in Russia's most remote and rugged region and no effort was made to launch a scientific expedition there. And then the 'tongue of flame' that engulfed the beautiful Siberian taiga was forgotten.

As the bright night skies disappeared from Europe, scientists' and newspapers' attention also moved from this strange phenomenon to other scientific and technological marvels of the year. The year's rich harvest includes Hermann Minkowski's definition of time as the 'fourth dimension'; Ernest Rutherford's detection of a single atom (Rutherford also won the 1908 Nobel Prize for Chemistry for his work in radioactivity); Herman Anschutz-Kampfe's invention of the gyroscope, a compass without a magnetic needle; Ikeda Kikunae's discovery of the food additive monosodium glutamate (MSG); Sullivan Thomas's invention of tea bags; the discovery of the first large deposit of petroleum in Persia (now Iran), marking the beginning of the Middle East oil boom; 'Count von Zeppelin and His Triumphant Airship' (the headline of a major feature article in *The New York Times* of 5 July 1908); and, of course, the introduction of Ford's assembly line that rolled out the Model T, the first mass-produced car (which came with Henry Ford's famous promise: 'Any color – so long as it's black.'

The year also had plenty to offer to those who liked their science a little bit spicy. There were stories of captains spooked by 'magnetic clouds' descending on their ships. The provincial American newspapers, as they still are, were full of reports of 'alien ships' racing at fantastic speeds. (The term 'flying saucer' was coined in June 1947, when an American pilot named Kenneth Arnold saw nine bright discs in the sky near Mount Rainier in Washington. He said the objects were moving like 'a saucer skipping across the water'.) All this paled into insignificance when the passengers on a steamship in the Gulf of Mexico described in detail the tale of a '200-feet long sea serpent' rising from the sea, which they had seen with their own eyes.

In the history of natural catastrophes, 1908 is remembered for two reasons: the Messina earthquake and the official introduction of Morse code SOS (... — ...) for the international signal of extreme distress. The most violent earthquake recorded in Europe's history killed 150,000 people in southern Italy and Sicily. The epicentre was Messina, Sicily's second biggest town. Had Messina been hit by the Tunguska fireball, the devastation would have been unimaginable. What if the fireball had hit a large city like St Petersburg or London? The sound of the SOS would still be echoing in humanity's ears.

Twenty years later

The first reports of the Tunguska fireball reached the Western world in 1928. The following year, C.J.P. Cave, a British astronomer, noticed the coincidence of the date of the fireball and the graphs recorded at six different locations in England on 30 June 1908. In 1930 another British astronomer, Francis John Welsh Whipple (not the American astronomer Fred Lawrence Whipple who, as we'll see, proposed the 'dirty snowball' model for comets), suggested that the airwaves in England were caused by the great Siberian meteorite, as the Tunguska fireball was known then. 'How it happened that the fall of the great meteor which produced the waves was not brought to the notice of the scientific world at the time is a mystery', he said at a meeting of the Royal Meteorological Society. 'There are many marvellous features in the story of the Siberian meteorite, a story without parallel in historical times. It is most remarkable that such an event should occur in our generation, and yet be so nearly ignored.'

Whipple used the six graphs to show that the pressure fronts had been travelling with an average speed of about 1,130 kilometres per hour. Pressure waves of similar intensity were recorded in Britain on 27 August 1883 when a volcanic eruption happened on the other side of the world, at Krakatoa on the Indonesian island of Rakata. Whipple also established that the first four 'almost similar' waves were registered within a period of two minutes, and then the final two waves followed. He concluded that there were two kinds of phenomena: the first four waves were caused by a meteorite passing through the atmosphere, and the last two waves referred to the meteorite striking the ground.

After reading Whipple's paper, another astronomer, Spencer Russell, immediately associated the remarkable night glows of 1908 with the Siberian meteorite. 'The entire northern sky on these two nights was of suffused red hue, varying from pink to an intense crimson', he recalled. 'There was a complete absence of scintillation or flickering, and no tendency for the formation of streamers, or a luminous arch, characteristic of auroral phenomena. Twilight on both of these nights was prolonged to daybreak, and there was no real darkness.'

Though they explained the cause of the phenomena observed in England, neither Whipple nor Spencer made any attempt to explain the nature of the great Siberian meteorite. This task was left to a relatively unknown young Russian scientist.

CHAPTER TWO

THE CASE OF A MISSING METEORITE

About 7 p.m., 16 June 1794, Siena, Italy. A vibrant town popular with English tourists. It has a population of nearly 30,000, and a university founded in 1240. Tourists sipping the local Chianti wine in Piazza del Campo, the town's large shell-shaped centre, notice a dark cloud spoiling a picture-perfect blue sky. As the cloud moves southeast they hear sounds of distant thunder.

A few kilometres away, villagers walking back to their houses in Cosona are bewildered as several stones hiss through the air and land at their feet. Soon afterwards a loud continuous sound, as if from artillery fire, fills the sky. A young woman named Lucrezia Scartelli is curious and picks up a stone the size of an olive, and immediately drops it as it burns her hand. She hears a thundering sound and sees another large stone falling. She runs away scared.

Later, another witness named Ferdinand Sguazzini tells other villagers that the stones came so fast that the big ones went right into the ground. As deep as my arms, he says, stretching his arms out. He then shows them a stone: it is the size of a tomato and dark black. He scratches it with his knife and his audience scream with surprise when they find it silvery white inside.

Stones from the heavens

Ambrogio Sodani, a professor of geology and zoology at the University of Siena at the time, studied the phenomenon and found that the stones fell in an area of 47 square kilometres. He estimated their number to be a few hundred. Their sizes ranged from a small pebble to a 3.5-kilogram rock. He tested them with a magnet and concluded that they were mostly iron. They appeared similar to other stones from the sky that he had seen before.

The famous biologist Lazzaro Spallanzani, who was in Naples in 1788 while Mount Vesuvius was in eruption, suggested that a tornado had carried the Siena stones from Vesuvius, 320 kilometres away, which had again erupted just eighteen hours earlier. But Sodani believed that the stones he had collected were different from the volcanic stones of Mount Vesuvius. He suggested that they had come from the sky. Everyone ignored his suggestion except Ernst Florens Friedrich Chladni, a physicist from Wittenberg in Germany (he was a corresponding member of the Academy of Sciences at St Petersburg, and rightly Russia also claims him).

Just two months before the stones fell in Siena, Chladni had published a slim book, *On the Origin of Iron-masses*, in Riga. In this book, he claimed that stones and masses of iron fall from the sky and some of them even create fireballs in the atmosphere. He suggested that these objects originated in 'cosmic space' and might be remnants of planet formation or planetary debris from explosions or collisions.

Chladni's idea that meteorites were extra-terrestrial in origin was scientific heresy: it was an attack on the great Newton himself, who believed that apart from the heavenly bodies – stars, planets and comets – all space beyond the moon was empty (the heavens are empty of all matter except a very thin, invisible ether, he said in 1704).

Chladni's book was ridiculed by the scientists of his time. 'By all means you must read Chladni's infamous book on iron masses', Alexander von Humboldt wrote to a friend. Georg C. Lichtenberg 'wished Chladni had not written his book'. He felt that Chladni had been 'hit on the head with one of his stones'.

The idea that rocks don't just fall out of the sky was so entrenched that even America's scientifically literate president, Thomas Jefferson, is believed to have commented, 'Gentlemen, I would rather believe that two Yankee professors would lie than believe that stones fall from heaven', when told that two Yale University professors had reported the fall of meteorites over Weston, Connecticut, in December 1807.

We do not know whether Jefferson's remark is truth or myth, but we do know for sure that the French Academy of Science was one of the staunchest critics of Chladni and continued to reject his ideas, even though stones were literally falling in front of bewildered witnesses (at Wold Cottage, England, on 13 December 1795; Evora, Portugal, on 19 February 1796; and Benares, India, on 19 December 1798). When a spectacular shower of several thousand stones fell near the town of L'Aigle in northern France on 26 April 1803, which was

21

witnessed by many French officials, the Academy hastily dispatched one of its members, physicist Jean Baptiste Biot, to investigate the phenomenon. 'I collected and compared the accounts of the inhabitants: at least I found some of the stones themselves on the spot, and they exhibited to me physical characters which admit of no doubt of the reality of their fall', Biot wrote in his report. The report finally convinced the scientific establishment that stones do fall from the heavens.

Astronomers now remember Chladni as the founder of meteoritics, the science of meteorites; physicists remember him as the founder of acoustics for his mathematical investigation of sound waves (the patterns formed when a thin plate, covered with sand, is made to vibrate are still called Chladni figures).

A lesson in meteoritics (and investing in metals)

Today, more than 200 years after the publication of Chladni's 'infamous' book, we know that meteorites (from the Greek word *meteoros*, meaning 'high in the air') are chunks of extra-terrestrial matter, remnants of geological processes that formed our solar system 4,600 million years ago. When these chunks enter Earth's atmosphere they shine brightly because of the heat produced by friction with the air. Most chunks are too small – usually the size of a grain of sand, but no larger than a pea – to survive the trip, and are called meteors (or falling stars or shooting stars because they leave momentary streaks of light in the sky).

Very rarely, a large chunk, which flashes like a fireball in the sky, survives its journey through the air to hit the ground. The falling object – a solid piece of stone or iron, often weighing many kilograms – is known as a meteorite. (Until a meteor or a meteorite enters Earth's atmosphere it is known as a meteoroid.) Asteroids (also called planetoids or minor planets), on the other hand, are small bodies orbiting the Sun, mostly in between Mars and Jupiter. Most meteorites are pieces of rock and/or metal from asteroids; most meteors are produced when comets disintegrate (comets are independent masses of ice and dust that orbit the Sun).

The fall of a large meteorite is a rare but spectacular event. The meteoroid enters Earth's atmosphere at a very high speed, ranging from 40,000 to 250,000 kilometres per hour. The friction against air not only decelerates it, but also raises its temperature. At a height of about 100 kilometres the meteoroid is so hot that it shines like a fireball. Its outer layer is continuously vaporised and ejected, leaving a trail of dust and smoke. At a height of about 20 kilometres, its speed has been so much reduced – it is about 10,000 kilometres per hour – that it no longer glows. It decelerates further to a free-fall speed of somewhere between 320 and 640 kilometres per hour until it reaches the ground. Because of high speed and high temperature, most large meteoroids break into several or sometimes thousands of fragments at a height of between 11 and 27 kilometres. The fragments strike the Earth in a roughly elliptical pattern a few kilometres long. The 'shower of stones' at Siena in 1794 was this type of fall.

A meteoroid's spectacular entry into the atmosphere is accompanied by an equally spectacular sonic boom. Because sound travels quite slowly, at only about 1,200 kilometres per hour, it takes from 30 seconds to several minutes after the appearance of the fireball before any sonic boom can be heard. However, many witnesses have claimed that they heard strange noises as a meteorite streaked across the sky. Known as electrophonic sounds, these range from hissing static to the sound of an express train travelling at high speed. Electrophonic sounds have not yet been validated scientifically, but scientists suspect that light given off by a meteorite must be accompanied by invisible electromagnetic radiation in the form of VLF (very low frequency) radio waves at frequencies from 10 hertz to 30 kilohertz. These waves could reach the observer as soon as the meteorite approached, but you wouldn't hear them. Often, the witness of such sound is located near metal objects. It is possible that such objects act like a transducer, converting inaudible electromagnetic waves into audible sound vibrations.

Meteorites contain various proportions of metals (iron and nickel) and stones (silicates). Thus, meteorites can be divided into three simple categories: irons consist mainly of metals; stones consist of silicates with little metal; and stony-irons contain abundant metals and silicates.

The largest known meteorite is still lying where it fell in prehistoric times in Hoba, Namibia. This room-sized meteorite is one metre high and weighs 60 tonnes. It is mostly iron. The most famous sacred meteorite is the black stone of the Ka'bah, which now lies in the Great

Mosque in Mecca, Saudi Arabia, towards which Muslims pray five times daily. Islamic tradition has it that the stone came from heaven and was originally hyacinth in colour before it turned black because of humanity's sins.

A large meteorite or asteroid hits the ground with such an enormous force that it shatters into pieces and leaves a big hole – a crater. How would you recognise a meteorite crater if you fell into one? When the meteorite shatters at the moment of impact, the pulverised earth and meteorite fragments are hurled out of the crater and scattered around it, but a considerable part falls back into it. This causes craters to display raised and overturned rims. But most of the time this above-surface evidence is erased as the Earth's surface is always changing. After thousands of years of weathering and erosion by wind, rain, ice, changes in temperature, gravity and activities of animals and plants, a crater may not look like a crater.

Perhaps the most important characteristic of a crater is the presence of meteorite fragments in the vicinity of it. Only small craters are expected to have meteorite fragments. If a meteorite explodes the moment it strikes the ground, most of it is changed into gas. The main feature of craters produced by such meteorites is the complete absence of fragments. Another feature is that these craters have diameters of more than 1 kilometre. Thus the absence of meteorite material is not evidence against a meteorite impact.

Even if a meteorite does not leave any fragments, it leaves some evidence of impact. Geologists look for three types of impact evidence:

25

- *Impactites*. The impact produces so much heat that the rocks melt and splatter into the air. As the drops of melted rock cool they turn into glassy globules, called impactites. They often contain iron-nickel grains, remnants of the meteorites. Depending upon how much glass and other minerals they contain, these globules are sometimes given specific names such as tektites, krystites and suevites.
- *Shocked quartz*. The impact also shatters the rocks, throwing tiny grains of quartz into the air. The shattering is so violent that it leaves patterns on these grains, known as shocked quartz.
- *Rare elements*. Certain elements such as iridium are rare in rocks in the Earth's crust. A gigantic impact can scatter these elements all over the world.

Since the late 1950s geologists have confirmed only 160 impact craters. They all were formed by metal meteorites, and their diameters range from 10 metres to 200 kilometres. No craters associated with stony meteorites have been found. The reason perhaps is that stony meteorites disintegrate in the atmosphere. Even if some fragments have survived the journey, they may not have survived the terrestrial weathering and erosion.

The world's first authenticated and best-preserved impact crater is in Arizona. Known simply as the Meteor Crater, its rim-to-rim diameter is 1.2 kilometres and its circumference is nearly 5 kilometres. Its depth below the surrounding plain is about 175 metres, with a 45-metre-high rim rising above the plain. It was gouged in about 50,000 years ago by a meteorite with the diameter about the width of a football field. The original meteorite, packed with more than 300,000 tonnes of iron and nickel, was travelling with a speed of about 64,000 kilometres per hour. It was strong enough to pass through the atmosphere without breaking into pieces.

The Meteor Crater has also found a place in the annals of foolhardy investments. In 1902 Daniel Moreau Barringer, a successful mining engineer from Philadelphia, heard about the crater and the small balls of iron scattered around it. He rejected the prevalent idea that the crater was formed by a volcano and formed the view that it was a meteorite crater. He estimated that an iron-nickel body weighing between 5 and 15 million tonnes lay beneath the surface. In 1903, without ever having seen the crater, he formed the Standard Iron Company and applied for and received from the United States government a 199-year lease (signed by President Theodore Roosevelt himself) on two square miles of land around the crater. Over the next 26 years, until Barringer's death in 1929, the Standard Iron Company spent more that $600,000 (a considerable fortune in those days) drilling scores of holes – the deepest reached 412 metres – but produced nothing except tiny samples of meteorite material which contained 93 per cent iron and 7 per cent nickel and traces of other elements, including precious platinum and iridium.

The 'fool's iron' probably gleamed in the eyes of Russian authorities when in 1921 they decided to find meteorite falls which had been recorded in Russia during and after the war years. The new regime formed after the October Revolution of 1917 (the Union of Soviet

Socialist Republics was born in 1922) was reeling economically when the news of the iron bonanza of the Meteor Crater reached Moscow. Authorities dreamed about discovering iron worth a fortune that had fallen within Russia. A bright 38-year-old scientist was put in charge of finding these 'treasures from space'.

Our man in Tunguska

Fast forward to 30 June 1958. The face that peers out from the 40-kopeck stamp released today by the USSR is that of a worried, bespectacled, grey-bearded man wearing a fur hat. The stamp commemorates the fiftieth anniversary of the Tunguska fireball and the scientist who devoted the last two decades of his life to solving the riddle of the mysterious Siberian meteorite. He was the first scientist to visit the Tunguska site.

From 1927, when he led the first expedition to Tunguska, to his death in 1942 in a Nazi prisoner-of-war camp, Leonid Alekseyevich Kulik, a mineralogist and an authority on meteorites, believed that the Tunguska fireball was a giant meteorite. His four expeditions to the site from 1927 to 1939 (the planned fifth expedition was postponed because of the outbreak of the Second World War) failed to find any remains of the meteorite, which he firmly believed had been lying hidden somewhere in the explosion site.

'Where was the meteorite crater with its raised rim that should have been created at the moment of impact?' The question haunted Kulik. The worried face in the photograph on the 1958 stamp seems to ask the question:

Figure 3: Leonid Alekseyevich Kulik, the first Tunguska researcher.
(Photo: Soviet Academy of Sciences.)

if it were not a giant meteorite, then what caused the great Siberian explosion? The mystery still eludes scientists. There are many theories, but no definitive answer. But first the man who made Tunguska famous.

Figure 4: The Soviet Union stamp honouring Kulik on the fiftieth anniversary of the Tunguska event in 1958.

The eldest son of a doctor, Kulik was born on 31 August 1883 in Tartu, Estonia. After graduating from high school with a gold medal, he studied at the St Petersburg Forestry Institute. In 1904 he was expelled for Bolshevik revolutionary activities, and spent the following year teaching mathematics and physics to adults in a night school. In 1906 he was imprisoned for a short time. After leaving prison, Kulik studied physics and mathematics at Kazan University. He married Lydia Ivanova in 1907, and his first daughter Helen was born in 1910 and his second daughter Irina in 1925.

In 1910 Kulik was again imprisoned for a short time and remained under police surveillance for two years. In 1912 he went to the Ural mountains where he worked as a forestry officer. There he met Vladimir I. Vernadsky, a highly respected geochemist and mineralogist, who was leading a group of geologists searching for mineral deposits. This chance meeting was the start of the metamorphosis of a revolutionary and an amateur poet into a pioneering scientist.

Vernadsky was so impressed with young Kulik's quick grasp of mineralogy that he predicted that this 'lover of minerals and nature' would one day become a major scientific researcher. Vernadsky arranged to have Kulik transferred from the forestry department to his own expedition, and eventually to the prestigious Mineralogical Museum of the Academy of Sciences at St Petersburg.

The start of the First Word War in 1914 put a stop to Kulik's rapid rise as a mineralogist. He was drafted into the Russian Army and fought in East Prussia. After the war he studied at the military academy and then continued working for the army as a scientist. During the October Revolution, Kulik moved to Tomsk, Siberia's major city, where he joined the Red Army and also taught mineralogy at Tomsk University.

After his discharge from the Red Army in 1920, Kulik returned to his museum post in St Petersburg. With the single-minded intensity that had characterised his life as a revolutionary, army officer and teacher, Kulik now applied himself to the study of meteorites and within a short period established himself as an authority on this relatively new branch of science. Also working at the museum at the time was Evgeniy L. Krinov, a highly respected mineralogist and an authority on meteorites. Krinov, who is best known for his authoritative book *Giant Meteorites* (published in English in 1966), called

Kulik 'a vibrant, cultured man around whom young people flocked'. Krinov also admired Kulik 'as an outspoken individual who was not afraid to voice his opinion when he was convinced he was right'. An admirable trait in any scientist.

The meteorite hunter

In 1921 the Soviet Academy of Sciences approved the country's first special meteorite expedition, which was charged with the task of locating and examining meteorites fallen in inhabited regions of Russia. One of the expedition's tasks was to gather information from local populations and talk to eyewitnesses. The expedition left Petrograd (as St Petersburg was known after the First World War; the old name sounded too German for contemporary Russians) on 5 September 1921 under the leadership of Kulik.

At this time Kulik was not aware of the Tunguska meteorite. He first heard about it at the railway station as the small expedition party set off on the Trans-Siberian Express. D.O. Sviatsky, editor of the magazine *Mirovedeniye*, ran up to the train and gave Kulik a page torn from the 1910 calendar published by Otto Kirchner of St Petersburg. On the back of this page was the following note:

About 8 a.m. in the middle of June 1908 a huge meteorite is said to have fallen in Tomsk, several sagenes [1 sagene = 2.314 metres] from the railway line near Filimonovo junction and less than 11 verst

[1 verst = 1.067 kilometres] from Kansk. Its fall was accompanied by a frightful roar and a deafening crash, which was heard more than 40 verst away. The passengers of a train approaching the junction at the time were struck by the unusual noise. The driver stopped the train and the passengers poured out to examine the fallen object, but they were unable to study the meteorite closely because it was red-hot. Later, when it had cooled, various men from the junction and engineers from the railway examined it, and probably dug around it. According to these people, the meteorite was almost entirely buried in the ground, and only the top of it protruded. It was a stone block, whitish in colour, and as much as 6 cubic sagenes in size.

Kulik was fascinated by the story, so fascinated that he immediately decided to investigate it further. Over the years his fascination would become an obsession that would consume the rest of his professional life.

The Academy of Sciences had provided Kulik's team with a carriage on the Trans-Siberian Express. They travelled across the Urals into Siberia, then made stops in Omsk, Tomsk and Krasnoyarsk, and finally arrived at Kansk. At Kansk, Kulik searched through Siberian newspapers published during the summer of 1908 for reports of a meteorite fall. He soon discovered that the calendar note was the beginning of an article published on 12 July 1908 in the newspaper *Sibirskaya Zhizn* from the Tomsk region. The article turned out to be wrong in almost every detail, except about the train stopping near Kansk.

As he sifted through newspapers he found many reports of a huge meteorite fall on the morning of 30 June 1908. He prepared a questionnaire and published it in local newspapers and distributed 2,500 copies among the locals. As a result, he collected breathtaking personal accounts, which were vivid and rich in details, from several dozen eyewitnesses who could still remember the event. From the information collected, Kulik painstakingly painted a picture of the meteorite, which he called the 'Filimonovo meteorite' (the term 'Tunguska meteorite' was not used until many years after the first Tunguska expedition in 1927).

Though he now firmly believed that between 5.00 and 8.00 a.m. on 30 June 1908 a giant meteorite flew in the general direction of south to north and fell probably in the basin of the Vanavara River, a tributary of the Stony Tunguska River, he was unable to embark upon a search. The expedition had run out of funds and the authorities needed his train carriage which had been lent for the expedition.

On return to Petrograd, Kulik submitted his report to the Academy of Sciences in which he suggested that the Siberian meteorite was a rare event in human history and must be investigated. His 'Account of the Meteorite Expedition' was published in the *Journal of the Soviet Academy of Sciences*. The Academy members, however, were sceptical of the claim. But Kulik was not alone in his quest.

A.V. Voznesensky, who was Director of the Irkutsk Magnetic and Meteorological Observatory in 1908, published a report in *Mirovedeniye* in August 1925 in which he claimed that the seismic and air waves recorded by his observatory on 30 June 1908 were both caused by the fall of a giant meteorite. He suggested that the air waves were caused by 'the explosion of the meteorite at a height of about 30 kilometres above the surface of Earth'. In his report, Voznesensky also rightly pointed out that his observatory's seismograph registration of the fall of the meteorite was the first in the history of science.

Voznesensky also suggested that the investigator of the spot where the Siberian meteorite fell would find something very similar to the meteorite crater of Arizona. 'The Indians of Arizona still preserve the legend that their ancestors saw a fiery chariot fall from the sky and penetrate the ground at the spot where the crater is; the present-day Evenki people have a similar legend about a new fiery stone', he said. He concluded his report with a tantalising idea: the search for the meteorite could prove a very profitable enterprise, particularly if this meteorite turned out to belong to the iron class.

In the same issue of *Mirovedeniye*, S.V. Obruchev, a geologist, wrote about his studies in the summer of 1924 in the Tunguska region. He also described stories of a huge calamity which had been told to him by the local indigenous inhabitants, the Tungus people (later named Evenki by the Soviets; they are probably the oldest surviving native Siberians). Obruchev speculated that the calamity was caused by a giant meteorite. 'In the eyes of the Tungus people, the meteorite is apparently sacred and they carefully conceal the place where it fell', he said. However, he learned that there was a 'flattened forest' three or four days northeast of Vanavara.

In 1926, I.M. Suslov, an ethnographer, visited the Tunguska region. In his report 'In Search of the Great Meteorite of 1908' in *Mirovedeniye* (March 1927), he described some of the 60 eyewitness accounts of the explosion he had collected. In these accounts, he said, such expressions were heard as 'the forest was crushed', 'the grain stores were destroyed', 'the reindeer were annihilated', 'people were injured', 'the taiga was flattened', and so forth. Suslov also visited the tent of Ilya Potapovich Petrov, the very same Evenki whom Obruchev had questioned in 1924. Ilya Potapovich, who would work as a guide on Kulik's 1927 expedition, agreed to Suslov's request to draw a map of the area of the catastrophe. 'Ilya Potapovich drew the map with coloured pencils, and a group of Tungus made corrections', Suslov said.

The articles of Voznesensky, Obruchev and Suslov convinced the Soviet Academy that an event of major importance had occurred and investigations should be continued. Kulik's mentor Vernadsky also supported his request. 'The expedition proposed by Kulik may turn out to have a very great scientific significance, and its results may repay a hundredfold the time and money spent on it', he wrote to the Academy. The Academy approved the first Tunguska expedition in 1926.

The first Tunguska expedition

In early February 1927 Kulik left Leningrad (as St Petersburg was again renamed in 1924) with one assistant named G.P. Gyulikh. Travelling by the Trans-Siberian

Express, on 12 February he reached the remote Siberian station of Taishet, some 900 kilometres south of the Tunguska explosion site. After buying food and supplies and other equipment, Kulik and his assistant left Taishet by horse-drawn sledges. Battling frequent snowstorms and bitter temperatures, it took them five days to reach the small village of Kezhma, about 215 kilometres south of the explosion site, where they replenished their food and supplies, and left with three carts on 22 March.

No one has better described inhospitable Siberia than the Russian writer Anton Chekhov. He was 30 years old when in 1890 he made an incredible trip, mainly by horse carriage and river boat, across Siberia to the island of Sakhalin, a penal colony in tsarist Russia. 'Why is it so cold in this Siberia of yours?' With this question of his coach driver, Chekhov's journal of his expedition, *A Journey to Sakhalin*, begins. 'Because that's the way God wants it', replies the driver. He travelled in relatively more inhabited areas of Siberia, yet Chekhov complained: 'Siberian highways have their scurvy little stations … They pop up every 20 or 25 miles. You drive at night, on and on, until you feel giddy and ill, but you keep on going, and if you venture to ask the driver, how many miles it is to the next station, he invariably says, "Not less than twelve".'

Kulik did not have the luxury of a coach or the opportunity to complain to someone. He and his assistant were now travelling through the most rugged taiga, split with creeks, gullies, bogs and swamps and steep hillsides. They had to take many detours to ford rivers because it was too dangerous to cross flimsy suspension bridges.

They were now in a 'vast and sinister' primeval forest in which 'the weak and imprudent often perish', as described by the Russian writer Yuri Semyonov in his book *The Conquest of Siberia* (1944). Nevertheless, three days later they arrived in Vanavara, the most northerly outpost of civilisation, a tiny trading station of only a few houses and stores, situated on the high right bank of the Stony Tunguska River.

Kulik had been carrying a letter from Suslov to the local Soviet political officer with a request to put Kulik in touch with the Evenki Ilya Potapovich. To Kulik's dismay, Ilya Potapovich flatly refused to guide him to the 'thunder god's home', the forbidden and sacred land. To the Evenki people, the fiery body was a visitation from Ogdy, their god of thunder, who had cursed the area by smashing trees and killing the animals. No one dared approach the site for fear of being cursed by Ogdy. Kulik's dogged determination eventually won over Ilya Potapovich's reluctance when he offered him two bags of flour, several rolls of cloth and building materials for the roof and floor of his house.

Kulik was keen to continue the journey immediately, and the day after arriving at Vanavara he set out with his assistant and new guide. Their horses were tired after the long journey, and being overloaded they could not make their way through the snow-covered forest. They were forced to return to Vanavara and await better weather.

On 8 April Kulik's party set off again. This time they were better prepared and their packhorses were loaded with enough food to last about a month. They travelled along the Stony Tunguska River for about 30 kilometres downstream until they reached the Chamba River. They then followed this river for about 10 kilometres and by nightfall reached the hut of the Evenki Okhchen, who agreed to join them as a second guide.

The next morning they reloaded all their supplies onto Okhchen's flock of ten reindeer and set off over a reindeer track along the Chamba River. Two days later the track came to an end. The five days' arduous journey had taken its toll. Exhausted and sick with scurvy and various infections from months of poor food, Kulik was still determined to go on. They hacked their way with axes through the virgin taiga. On 13 April the expedition crossed the Makrita River, where they found the beginning of a mass of fallen trees, uprooted as by an explosion. In the distance could be seen the twin-peaked mountain called Shakrama by the Evenki people. On 15 April Kulik climbed the mountain and saw the explosion site stretching to the horizon before him. 'This is where the thunder and lightning fell down', pointed out Ilya Potapovich, 'and burned down my relative Onkoul's grain store'. (The remains of the grain store were indeed discovered by Kulik's third expedition.)

Kulik saw an oval plateau 70 kilometres wide where the forest had been flattened, all the trees stripped and snapped off in the direction of the blast. 'The results of even a cursory examination exceeded all the tales of the eyewitnesses and my wildest expectation', Kulik wrote in his diary. 'One has an uncanny feeling when one sees 50- to 75-centimetre thick giant trees snapped across like twigs, and their tops hurled many metres away to the south.'

Figure 5: Charred and fallen trees near the blast site as seen by Kulik. (Photo by N.A. Strukov, Moscow, 1928.)

Kulik wanted to explore the centre of the blast area, which he assumed lay beyond the distant snow-covered ridges to the north where the forest had been completely destroyed, but his Evenki guides were extremely superstitious and refused to walk through the taiga burned by their god Ogdy. He had no choice but to return to Vanavara.

Kulik was exhausted but determined to find the fall point. Back in Vanavara on 22 April, he hired Russian peasants from Kezhma village and planned a new route to the explosion site. The expedition left Vanavara on 30 April. After three days' journey by sledge they again reached the Chamba River. This time Kulik decided to build rafts and navigate first up the Chamba and then along the Khushmo River, which were flooded with the thawing snow. Ten days later the expedition reached the mouth of the Churgima River, a tributary of the Khushmo. On 20 May they finally arrived at the edge of the devastated taiga. Kulik decided to camp there.

The next day as he followed the direction of the fallen trees for a few kilometres he reached a marshy basin between 5 and 7 kilometres in diameter and surrounded by low-lying hills. In their legends the Evenki people referred to the area as the South Swamp, but to Kulik it resembled a gigantic cauldron and he named it the Great Cauldron. Here the devastation was greater than what he had seen from Mount Shakrama. Kulik decided to transfer his camp there. Over the next few days he walked around the hills, climbed them and measured the direction of fallen trees. Kulik was now convinced that he had found the epicentre of the fall. He wrote later: 'There could be no doubt. I had circled the centre of the fall. With a fiery stream of hot gases and cold solid bodies, the meteorite had struck the cauldron, with its hills, tundra and swamp.'

Everywhere, for a distance of more than 30 kilometres from the centre, was like a forest of 'telegraph poles', dead trees still standing, but their twigs and branches blown away. 'The taiga has been practically destroyed by being completely flattened', he recorded in his diary. 'The trees lie in rows on the ground, without branches or bark, in the direction opposite to the centre of the fall. This peculiar "fan" pattern of fallen trees can be seen very well from some of the heights that form the peripheral ring of trees.'

41

Figure 6: The forest of 'telegraph poles' as seen by Kulik. (Photo by I.M. Suslov, Moscow, 1928.)

There was another remarkable feature: within the central blasted area was a ring of upright trees, completely stripped of foliage. The fact that they had remained upright while all trees outside the ring had been flattened, Kulik thought, marked some kind of node or region of rest where air waves cancelled each other.

There was also the evidence of fire; some of the trees were charred, but this evidence of burning was unusual: in a forest fire, trees are usually burnt on the lower part of their trunks, but these had been burnt uniformly and continuously. Kulik believed that a great rush of hot air produced by the change of kinetic energy into heat energy when the meteorite crashed into the Earth blew the trees down and scorched them.

In some areas Kulik also found forest growth about twenty years old. 'From our observation point no sign of forest can be seen, for everything has been devastated and burned, and around the edges of this dead area the young twenty-year-old forest growth has moved forward furiously, seeking sunshine and life', he wrote in his diary.

Kulik also noticed circular giant ridges, like waves in water, which he believed were formed when the solid ground heaved outwards under the impact of the meteorite. The whole scene was like a giant picture of what happens when a brick from a wall falls into a puddle of mud. Kulik had expected to find the evidence of a giant meteorite in the central part of the basin, but found that the area was dotted with dozens of holes 'exactly like lunar craters'. These funnel-shaped holes ranged from 10 to 50 metres across and up to 4 metres deep. Their edges were mostly steep, the bottoms flat and swampy.

'I cannot say how deeply the meteorites had gone into the tundra and the rocks', he wrote in his account of the expedition, *Beyond the Tunguska Meteorite*. 'It was impossible for me to go right round the whole area ... or do any digging. We had food left for only three or four days, our road was a long one and our one thought now was to get back safely. It was flight in the full sense of the word.'

He arrived at Vanavara at the end of June after nine days of travelling. There followed another three weeks on a raft on the Stony Tunguska River to the town of Yenisei, and then a comfortable journey by steamship to Krasnoyarsk and by train to Leningrad. Kulik was already dreaming about his next expedition.

In his report to the Academy of Sciences he wrote:

'This picture [of shallow holes] corresponds exactly to the theoretical conditions of fall of a swarm of large meteorite fragments, the larger specimens of which exceeded 130 tons. In all probability these fragments were of iron meteorites ... giant craters such as the Arizona crater are strewn with fragments of iron meteorite.' He concluded his report by pointing out: 'Since this fall occurred on the territory of the Soviet Union we are duty bound to study it.'

On 13 March 1928 the 'duty bound' Academy approved the second expedition with the aim of continuing the study of the Tunguska meteorite. The Academy seemed to agree with Kulik's exhortation in his report: the significance of the Tunguska fall 'will be fully appreciated only in history and it is necessary to record all the remaining traces of this fall for posterity'. However, the Academy granted limited funding that allowed only for mapping the blast area and magnetic survey of the holes. Kulik was also expected to recover meteorite fragments for the Mineralogical Museum.

Kulik's first expedition also attracted the attention of the Western press. In a detailed scientific article entitled 'The Great Siberian Meteorite: An Account of the Most Remarkable Astronomical Event of the Twentieth Century', in *Scientific American* (July 1928), Chas P. Olivier of the International Astronomical Association wrote: 'Fortunately for humanity, this meteoric fall happened in a region where there were no inhabitants ... but if such a thing can happen in Siberia there is no known reason why the same could not happen in the United States.' *The Literary Digest* of 30 June 1928 also

44

warned: 'Had chance directed this enormous visitor from space to the site of a city or a thickly settled country the world would have experienced an unparalleled disaster; one, we must not forget, which may yet happen should another such meteorite ever arrive.' After more than three-quarters of a century, these doomsday warnings remain 'current' – just swap the word 'meteorite' with 'asteroid'.

The second Tunguska expedition

In April 1928 Kulik left Leningrad on his second expedition to the Tunguska region. He was accompanied by his assistant V. Sytin, a hunter and zoologist. Kulik did not have any experts from other disciplines of science to study the explosion site. At Vanavara he was joined by a cinematographer, N.A. Strukov, from Moscow's Sovkino Studio.

Spring floods delayed the expedition's progress and the group, including five workers, reached the explosion site in late June. Kulik set up his camp and started his investigations. He surveyed an area of 100 square kilometres and marked 150 craters with wooden stakes. He also tried to dig into two of the craters but the water and the boggy soil made digging impossible. He did not have any pumps to drain water from the holes. His primitive magnetic instruments failed to detect any metal pieces 'brighter than the blade of a knife and resembling in colour a silver coin' which some Evenkis had reported finding in the devastated forest.

After a few weeks Strukov left, accompanied by three

workers. Kulik, Sytin and two workers remained. Kulik continued to collect samples of peat and other plant materials for microscopic examination back in Leningrad. By the beginning of August, Sytin and two workers were showing signs of vitamin deficiency. Kulik was also running out of funds. His prospects of finding the meteorite were dim. Kulik found himself in a dilemma. He knew if he went back without any results, his opponents in the Academy would deny him any funds for future expeditions. He worked out a strategy: he would stay behind but would send two workers to Vanavara and Sytin to Moscow to convince the Academy to approve more funding.

Sytin's arrival in Moscow coincided with the dramatic rescue of the crew of the airship *Italia*, which had crashed near the North Pole, by the Soviet ice-breaker *Krasin*. When Moscow newspapers got hold of Sytin's story of a Russian scientist who was risking his life in the wilds of Siberia to find a mysterious visitor from space, and whose exhausted and sick assistant had come to Moscow to seek funds, they had found another sensational story of adventure.

The Academy of Sciences buckled under public opinion. Funding was immediately approved for a rescue expedition and to continue investigation. Sytin arrived back at the explosion site accompanied by ethnographer Suslov and a group of journalists. Suslov had never visited the place he had written about in *Mirovedeniye*. Kulik was so pleased to see Suslov that he named the largest hole the 'Suslov crater'. Kulik quickly put everyone to work helping with magnetic

Figure 7: A recent photograph of the Suslov crater. (Photo by Vitalii Romeiko, Moscow.)

measurements of various holes. He started with the 50-metre Suslov crater. With the journalists watching, he hoped to make a discovery, but was disappointed not to find any trace of metal in the hole.

At the end of October the expedition returned to Vanavara. When he arrived in Leningrad at the end of November, Kulik was a national hero. He didn't need spinmeisters to spin the Tunguska meteorite story. Strukov's short film of his journey, *In Search of the Tunguska Meteorite*, was also a big help in advancing Kulik's cause.

Kulik's second expedition was widely reported in British and American newspapers. The London *Times* of 26 November 1928 noted that Professor Kulik 'has reached Krasonyars … having tolerably recovered from the hardships of his journey'. *The New York Times* reported on 2 December: 'Professor Leonid Kulik, Russian geologist, is now reported to be on his way to Leningrad from the depths of North-eastern Siberia, where a relief expedition found him two months ago after he had been given for dead.' A phrase in the report's long headline also hinted at the treasures buried in meteorite craters: 'METEORIC IRON DEPOSITS ESTIMATED AT $1,000,000'. *The Literary Digest* of 16 March 1929 published an account of an interview with Kulik and Sytin, which concluded with the following comment: 'The value of the metals in the Siberian find is estimated by Mr Sitin [*sic*] as between one hundred million and two hundred million dollars, chiefly for the iron and platinum … the chief object of further investigation of the site … is not the recovery of any valuable materials that may exist, but the obtaining of further scientific information about an event almost unique in the recorded history.' It seems that the stories of Barringer's holes in the Meteor Crater were still uppermost in people's minds.

Kulik was an excellent writer and speaker and made meteorites popular among the Soviet population. He made his Moscow audience shiver when, in a lecture, accompanied by Strukov's 'moving pictures of the appalling desolation', he remarked: 'Thus, had this meteorite fallen in central Belgium, there would have been no living creature left in the whole country; on London, none left alive south of Manchester or east of Bristol. Had it fallen on New York, Philadelphia might have escaped with only its windows shattered, and New Haven and Boston escaped too. But all life in the central area of the meteor's impact would have been blotted out instantaneously.'

There were critics as well. A few geologists continued to voice their doubts about a meteoritic origin of the Tunguska blast. They explained Kulik's 'craters' as the result of permafrost and pointed out that similar holes are often found in other parts of Siberia. Nevertheless, the Academy of Sciences approved the third expedition under the leadership of Kulik.

The third Tunguska expedition

The third expedition left Leningrad in February 1929 and returned in October 1930. It was a much larger undertaking which included many scientists. Among

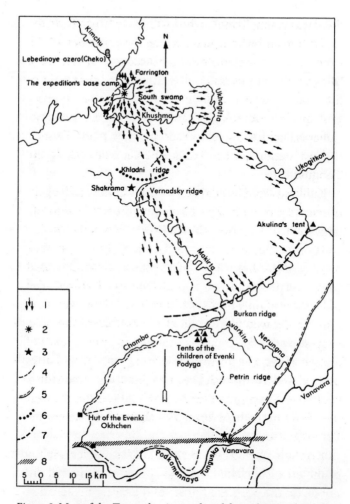

Figure 8: Map of the Tunguska site produced from the visual survey during Kulik's third expedition in 1929–30. (Key: 1 – devastated forest; 2 – meteorite impact site; 3 – survey points; 4 – track; 5 – road to Strelka trading station; 6 – limit of scorched area; 7 – limit of forest devastation; 8 – limit of the effect of the explosive wave.) (Courtesy Marek Zbik, University of South Australia.)

them was young Krinov (who lost a toe to frostbite on the expedition and who would become the Chairman of the Committee on Meteorites of the Academy of Sciences). The expedition was also well equipped. A horse train of fifty carts carried drilling machines, water pumps, geological, meteorological and surveying instruments, cameras, field tools and other supplies from Taishet, the last station on the Trans-Siberian railways, to the Tunguska site.

Kulik was convinced that the Suslov crater and a chain of craters around it were formed by the fall of separate large meteorite pieces. He decided to dig the Suslov crater first, his team drilling a 4-metre hole in it. The digging of this hole took Kulik's men one month's hard work, but they did not find any impact features. The walls showed only undisturbed material. However, they found a decayed tree stump at the bottom of the hole, suggesting that the crater was not caused by the impact of a piece of the meteorite after it broke on hitting Earth. 'Indeed, it was impossible to imagine that a tree stump could have been presented in a natural position so near the centre of the hole formed by the fall of a meteorite', Krinov noted.

Krinov, who had been making an independent survey of the area, concluded that the epicentre of the fall did not lie on the northern border of the basin, the location of the Suslov crater, as believed by Kulik. He suggested that the exact location of the epicentre was South Swamp, an area a few kilometres south of the Suslov crater. Krinov's suggestion made Kulik angry and he excluded him from further work on the expedition.

Kulik's belief in his Suslov crater hypothesis was unshakeable. He directed his men to set the drilling rig at the edge of the crater. They continued digging during the frigid winter months and drilled one hole 34 metres deep and 4 metres wide in the solid frozen ground. Still they did not find any meteorite material. Kulik's spirits soared when a worker found a piece of glass in the hole. Kulik concluded that it was an impactite, a rock fused into glass by the impact of the meteorite. But the glass did not turn out to be the smoking gun Kulik was looking for. It was, in fact, a piece of a bottle shattered during a fire which broke out in one of the huts on the first night of the expedition. 'Unfortunately this fragment was mentioned in several articles by Kulik as a find of silica glass, and even today it misleads researchers', Krinov said.

Two more holes were drilled before Kulik gave up drilling on 1 March 1930. He finally concluded that the Suslov crater was not a meteorite crater. Kulik learned the hard way what is now known to every student of meteoritics: finding a meteorite crater may not be as simple as finding a hole in the ground.

Six months later Kulik returned to Leningrad 'with grey hair and ruined health'. J.G. Crowther, a British science journalist who interviewed Kulik after the third expedition, wrote in *Scientific American* (May 1931): 'Professor Kulik's expeditions have left a mark on him. He is a tall, wiry, bronzed man of Scots figure, lean and a little tired. Perhaps a rest will soon entirely refresh him.'

Crowther's hope turned out to be a prophesy. Kulik was not to return to Tunguska for another seven years, but he continued working on the project. He also apologised to Krinov and requested him to continue working with him. Like Krinov, he now believed that the South Swamp was indeed the blast's centre.

Solving the riddle

In 1938 an aerial photographic survey of the Tunguska region was undertaken. Although the survey was incomplete, it showed the unique radial nature of the fallen trees. It also showed that the South Swamp was indeed the centre of some great catastrophe.

In July 1939 a 'refreshed' Kulik again returned to Tunguska to examine the South Swamp. His team started boring in several places in the swamp. The drilling showed numerous channels on the surface of the swamp. Kulik interpreted them as underwater craters. This idea was later rejected by geologists who said that this was a natural feature of the swamps in the area. However, the Academy of Sciences congratulated Kulik for his 'great persistence and enthusiasm that led to the recent concrete advance in our knowledge of the subject' and approved a fifth expedition in 1940 to conduct a magnetic survey of the South Swamp. But the expedition did not take place because of the outbreak of the Second World War.

On 5 July 1941, the day the Nazis invaded Russia, Kulik joined the Moscow People's Militia, a volunteer military unit. In October, while taking part in a battle on the front line, he was wounded in the leg and captured by the advancing German army. He was held as a prisoner of war in the town of Spas-Demensk, about 300 kilometres southwest of Moscow. He worked as a nurse in a prisoner-of-war hospital, where he contracted typhus and died on 14 April 1942. He was 58 years old. He was buried in the local cemetery. In 1960 the Academy of Sciences built a simple memorial on his grave. The plain gravestone is still there; it's simply marked: 'Kulik, Leonid Alekseyevich 1883–1942'.

Figure 9: Kulik's grave in Spas-Demensk, Russia.
(Photo by Andrei Ol'khovatov, Moscow.)

'Kulik', a 58-metre-diameter crater on the far side of the Moon, asteroid '2794 Kulik' and a Vanavara street named after him perpetuate the memory of the first Tunguska researcher and the founder of meteoritics in Russia.

The first expedition after the war was in 1958. In 1963 investigations gained new vigour under the leadership of Nikolai Vasilyev (1930–2001) of the Academy of Sciences, who coordinated the scientific research of 29 investigations. It wasn't until 1989 that foreign scientists were officially invited to join the Russian investigations. Now the Russian government has set aside 4,000 square kilometres of the Tunguska region as a national reserve. But the Tunguska explosion site remains inaccessible. The nearest Trans-Siberian railway station is Krasnoyarsk, 600 kilometres north of Vanavara. This tiny trading post has now grown into a small town of more than 4,000 people. From Vanavara the explosion site is about 70 kilometres, but it can be reached only by helicopter or by hiking.

Why did Kulik fail to find any meteorite fragments or impact crater in the South Swamp or the Great Cauldron? According to Krinov, careful investigation of the cauldron 'does not give any grounds for concluding that this cauldron is the place where the meteorite fell'. But four observations point to the fact that the cauldron is the site of the explosion: (1) the absence of other places in the Tunguska area which might attract attention as the possible place of the fall; (2) the Evenki people's designation of the cauldron as the place of the fall; (3) the cauldron is the epicentre of the seismic wave; and (4) the radial forest devastation around the cauldron. 'There is only one possible explanation that removes the contradiction, that is, that the meteorite did not explode on the surface of the ground, but in the air at a certain height above the cauldron', Krinov concluded.

Krinov's explanation did not solve the riddle of the Great Siberian explosion. The controversy about the Tunguska fireball continues to this day, and there is no shortage of attempts to explain the cataclysmic explosion. Igor Zotkin, a Russian expert on meteorites, once remarked: 'I doubt if there is any recent discovery that has not been called on to explain the Tunguska enigma.' Today, scientists' line-up of suspects includes a comet, a mini black hole, an asteroid, a rock of anti-matter or a mirror matter asteroid, and a methane gas blast from below. In the X-files we have an alien spacecraft, a laser beam fired by ETs in an attempt to communicate with lonely little earthlings, and an experiment on a 'death ray' which got out of hand.

CHAPTER THREE

THE TALE OF A
FIERY COMET

In the 1880s H.H. Warner, a wealthy American renowned as the 'patent medicine king', awarded a cash prize of $200 to an American or Canadian discoverer of a comet. The prize motivated Edward Emerson Barnard, a young amateur astronomer with virtually no formal education, to discover eight comets within six years. He won enough money to build himself a house. 'This fact goes to prove the great error', he said, 'of those scientific men who figure out that a comet is but a flimsy affair after all – for here was a house, albeit a small one, built entirely out of them. True, it took several good-sized comets to do it, but it was done nevertheless.'

Barnard's discoveries also earned him a scholarship to Vanderbilt University. In 1887 the 30-year-old graduate joined the Lick Observatory of the University of California at Santa Cruz (in his resumé he listed all his discoveries – ten comets and 23 nebulae – as well as his good habits: 'I am perfectly temperate, neither smoke, chew, nor use intoxicating drinks.'). Barnard, 'the man who was never known to sleep', was well known as an inexhaustible observer of the heavens when he became the subject of an elaborate hoax.

On 8 March 1891, when he opened his copy of the

San Francisco Examiner, Barnard was astonished and horrified to read a story describing his 'invention of a machine for scanning the skies and catching wandering comets on the photographic plate':

DISCOVERS COMETS ALL BY ITSELF
THE METEOR GETS IN RANGE, 'ELECTRICITY DOES THE REST.'
A Wonderful Scientific Invention that will do away with the
Astronomer's Weary Hours of Searching—The Idea Founded
on the Spectrum of the Comet's Light—It's Just Like
Gunning for Wandering Stars with a Telescope.

The long, breezy headline was followed by an equally long and playful story that filled two whole columns of the newspaper and included three detailed illustrations: a view of the complete comet-seeker, a diagram of an objective prism and an electrical circuit. The story also included several quotes from Barnard, whom the story described as 'the renowned young astronomer'. 'When the comet is caught, as in a trap … An alarm-bell rings in my bedroom down at the cottage', Barnard was quoted as saying. 'Of course, the signal quickly summons me to the roof [of the observatory] … A single glance should suffice to reveal the position of the new comet.'

Barnard immediately sent angry letters of denial to all San Francisco newspapers, but they all ignored them. Somehow the hoaxer had convinced the newspapers to ignore Barnard if he 'disowns his invention'. For two years, until the *Examiner* apologised in an editorial after his discovery of the fifth moon of Jupiter, Barnard continued to receive letters from all over the world. Even the famous astronomer Lewis Swift wrote that he had read the article 'regarding your invention to search for comets while asleep or using the 12-inch or playing poker … although the article appears somewhat fishy I am inclined to think it is still another of the marvellous inventions of the 19th century'.

Barnard never discovered the identity of the perpetrator of the hoax; however, he suspected one of his colleagues. Barnard died in 1923 leaving an astonishing legacy of observations – of planets, satellites, comets, double stars, bright and dark nebulae and globular clusters – that make him one of the greatest observers of all time.

Carolyn's comet

Barnard was twenty years old when he discovered his first comet in 1881; Carolyn Shoemaker was 54 when she discovered her first comet in 1983. Though she doesn't collect a 'cash prize of $200' for each of her discoveries, she has now 32 comets to her name, more than any other astronomer, living or dead (Jean-Louis Pons, a 19th-century amateur astronomer, discovered 37 comets but only 26 bear his name).

After spending 25 years as 'homemaker and mother' to her three children, Carolyn joined her better-known husband Eugene in 1980 in his search for comets and asteroids. Eugene, a geologist who is considered the father of planetary impact geology and has 29 comets to his name, was killed in a car accident in 1997 while hunting for impact craters in outback Australia.

Carolyn shares her most famous comet discovery with her husband and David Levy, an amateur astronomer who has 21 comets to his name, thirteen with Eugene and Carolyn Shoemaker. On 25 March 1993 Carolyn was as usual scanning films of the sky taken the previous night by Eugene and David from the 18-inch Schmidt telescope at Palomar Observatory in California. Unlike the observatory's main 200-inch Hale telescope which can see deep into space, the old Schmidt is an ideal telescope for surveying a wide section of sky.

There is still no 'automatic comet-seeker' to help Carolyn. She spends long hours in a dark room scrutinising pairs of photographs of the same area of sky with a stereomicroscope. The pairs of photographs are taken 45 minutes apart. As the comet will move across the sky in 45 minutes, it will be in a different position on the second film. The stereomicroscope allows Carolyn to see the two films simultaneously – one by the left eye and the other by the right eye. When viewed this way, a comet would appear to 'float' above the flat surface of the fixed stars. But it is not as simple as it sounds: a speck of dust, a satellite or a spark of light in the telescope can make comet-hunting a painstakingly slow sport (Levy considers it a 'competitive sport').

As she was slowly moving the film through the stereo-microscope she saw a blob that looked like 'a squashed comet'. When Eugene and David looked at it, they were surprised. It was the strangest thing any of them had ever seen during all their collective years of comet hunting. They had just discovered the comet of the century.

Following the tradition, the comet was named after its

discoverers – Periodic Comet Shoemaker-Levy 9 1993e (it was the ninth comet discovered jointly by the Shoemakers and Levy which travelled around the Sun in a short-period orbit; and the fifth comet discovered in 1993, hence 'e', the fifth letter).

When other astronomers trained their telescopes on this curious find, they noticed that Shoemaker-Levy 9, or S-L 9 for short, was in fact a string of 21 comet fragments stretched out in a trail nearly 200,000 kilometres long. S-L 9 had two unusual features. First, it had 21 separate nuclei like pearls on a necklace; no comet observed had broken into as many pieces. Many comets break up when they come close to the Sun, but they usually break into two or three pieces. Second, comets usually orbit the Sun; S-L 9 was orbiting Jupiter.

Astronomers also calculated that S-L 9 would inevitably crash into Jupiter in July 1994. Carolyn Shoemaker is not used to losing her comets. 'If I am going to lose a comet, then I want it to go out with fireworks', she hoped. That's just what happened: S-L 9 died in an extraordinary cosmic firework, which Carolyn watched from a safe distance of a billion kilometres.

By Jove, yes, it was the show of shows

In July 1994 the Hubble Space Telescope, orbiting 600 kilometres above Earth, and hundreds of thousands of telescopes around the world were aimed at Jupiter to watch the celestial drama of the century. Never in history had anyone witnessed the cataclysmic collision of two worlds, a comet crashing into a planet.

A few days before the crash, the observers noticed a trail of 21 comet pieces stretching for a length of more than 5 million kilometres, more than twelve times the distance between the Earth and the Moon. On 16 July 1994, the first piece with a width of 1 kilometre – the puniest in the line-up – slammed into Jupiter. Over the next six days twenty more pieces hit the giant planet.

When plunged into Jupiter's visible clouds each piece of S-L 9 was travelling with a speed of 200,000 kilometres per hour. As the piece entered the denser atmosphere it flattened like a pancake and then disintegrated, dumping energy equivalent to a blast of more than a million megatons (explosive energy is measured in kilotons or megatons; 1 kiloton equals 1,000 tonnes and 1 megaton 1 million tonnes of high explosive TNT). Taken all together, the 21 fragments released about 40 million megatons of explosive energy. Compare this energy with the Hiroshima atomic bomb: the mere 15-kiloton blast killed 140,000 people, injured hundreds of thousands and destroyed 70,000 buildings. The largest hydrogen bomb ever tested was a 58-megaton Russian bomb in 1961. A 40-million-megaton blast defies comprehension.

The explosion expelled the atmosphere into bright plumes of gas and debris, which at their peak towered 2,200 kilometres above the clouds. After a few minutes the plume plunged back into the atmosphere. The collapsing plume and the underlying atmosphere became hot and started releasing great fireballs of infrared rays. The point of impact turned into a dark scar, dubbed 'black eye' by astronomers, thousands of kilometres across.

This scary scenario was repeated twenty times, every seven hours, on average. The 'string of pearls' had been reduced to a chain of smouldering scars girdling the planet. Some fragments created bigger bangs than others. The largest of the fragments, a solid body 3 kilometres across, left a 'black eye' of debris as large as the Earth. Some of the black eyes took months to fade from Jupiter's face.

The diameter of Jupiter is eleven times greater than that of Earth; a piece of S-L 9 was only about a kilometre wide. Had it struck Earth it would have gouged a crater 60 kilometres across. Comparing S-L 9 with Jupiter is like comparing a fly with a dinosaur. If such a puny crash could cause so much damage, what would happen if a large comet hit Earth?

Did a comet strike Earth in 1908?

The Tunguska fireball was not a meteorite but a small gaseous comet which had left no trace of itself after the impact. That's what two astronomers concluded independently in 1934. F.J.W. Whipple, head of the Kew Observatory in London, said that in view of the fact that the recorded observations of the phenomenon such as bright nights and airwaves were confined to the north of Europe, it was suggested that 'the meteor was essentially a small comet and that the tail of the comet was caught by the atmosphere'. However, he was quick to admit: 'I do not feel much confidence in this hypothesis.'

I.S. Astapowitsch of Leningrad State University expanded the idea suggested by Academician Vladimir

Vernadsky, Kulik's mentor, that the Tunguska meteorite could have been a rather dense cloud of cosmic dust, possibly associated with a comet.

Astapowitsch asserted that the bright atmospheric phenomena following the blast could have been caused by the dust tail of the nucleus of a small comet rushing towards Earth and exploding with a tremendous energy. 'The explosion gave rise to seismic and airwaves', he said, 'while the high-temperature explosive waves caused the uniform scorching over the first few kilometres'. He also estimated the blast's energy to be between 1 and 2 megatons, about 100 times higher than Whipple's calculation (current estimates vary from 10 to 20 megatons).

In 1942 William H. Christie of the Mount Wilson Observatory in California analysed all available data on the Tunguska explosion and concluded that the data agreed with the action of a comet. 'The night glows, I think, remove all the doubts as to the nature of the object which struck Earth', he said. 'It was, apparently, a small comet ... and the tail, or part of it, which was captured by Earth formed the glows. The Greenwich time of the collision was just past midnight, hence no glows were observed much west of Greenwich because this hemisphere would be on the side away from the comet.' The Siberian blast appeared to be the only recorded instance of the collision of Earth with a comet, he said. 'How long we will have to wait for another such visitor we do not know, but let us hope it again chooses a sparsely inhabited region of this globe of ours for its final resting place?'

At that time scientists knew very little about the structure of nuclei of comets. They believed them to consist of one or a few large stony rocks or even a 'sandbag' of small particles. The size of nuclei was also over-estimated to be several hundred kilometres. Astapowitsch and Whipple's comet hypothesis also turned out to be as questionable as Kulik's meteorite hypothesis, and it was soon dumped by scientists. It was revived in the late 1950s, when the American astronomer Fred Whipple proposed his 'dirty snowball' model for comets.

A comet is a unique cosmic phenomenon: it suddenly appears in the sky, it blazes for a few days, it wows earthlings, it disappears. A comet's awe-inspiring spectacle has always intrigued people. Its irregular appearance in the sky, its varying size, form and brightness, its exotic tail, its abrupt disappearance – these were the mysteries which ancient people were unable to solve. To them comets were omens of disaster. According to the Greek poet Homer, who lived in the 8th century BC, a comet (from the Greek word *kome*, meaning 'hair') was 'a red star that from his flaming hair shakes down disease, pestilence and war'. In Shakespeare's *Julius Caesar*, after seeing a comet, Caesar's wife Calpurnia warns him: 'When beggars die, there are no comets seen; the heavens themselves blaze forth the death of princes.' Comets are no longer considered harbingers of doom, but they still intrigue astronomers.

Even now, astronomers know very little about comets, and they can't predict when the next one will come. Comets are chunks of matter left over from the birth of the solar system 4,600 million years ago. There are somewhere between 2 trillion and 5 trillion comets that circle the solar system in a halo-like cloud – the Oort cloud – between 20,000 and 100,000 astronomical units from the Sun (one astronomical unit is the distance between the Earth and Sun, about 150 million kilometres). In the Oort cloud, comets are not packed like sardines – the neighbouring comets are typically tens of millions of kilometres apart. The Oort cloud, sometimes called the Siberia of comets because of its freezing temperatures as low as –270 degrees Celsius, is named after Dutch astronomer Jan H. Oort, who in 1950 suggested the existence of a spherical reservoir of comets swirling around the solar system.

At about the same time, Gerard P. Kuiper of the University of Chicago suggested that there is another reservoir of comets, now known as the Kuiper belt. The Kuiper belt extends between 35 and a few hundred astronomical units from the Sun, beyond the orbit of Neptune, and contains some 200 million comets waiting in dusty cold storage for a chance to whoosh across our skies and strike terror into the heart of earthlings.

About a dozen stars pass within about 200,000 astronomical units of the Sun every one million years. Occasionally these passing stars push a comet out of its orbit, sending it towards the inner solar system. It may pass the solar system once and never return again, or it may settle into an orbit to visit us regularly, like Halley's comet. Comets that take twenty or more years to orbit the Sun, including Halley's, come from the Oort cloud; comets with shorter orbital periods start their journey towards the Sun from the Kuiper belt. Orbits of some comets have periods exceeding 1 million years.

Balls of ice and dust

Comets are fossils – frozen relics from the time of the infant Sun. By studying them, astronomers can find out how the Sun and the planets were born. In the early 1950s Fred Whipple and other astronomers provided an insight into the structure of comets. Whipple said that a typical comet has three parts: a frozen central part called a nucleus, a fuzzy cloud surrounding the nucleus called a coma (or head), and a tail consisting of gas and dust. The nucleus, usually only a few kilometres across, is a 'dirty snowball' made of grains of frozen mass consisting of water, methane, ethane, carbon dioxide, ammonia and many other gases. In 1986 the European Space Agency's *Giotto* spacecraft proved that Whipple's model was fairly accurate when it took close-up photographs (from a distance of 480 kilometres) of the nucleus of Halley's comet: a comet nucleus resembles a fluffy snowball coated with a crust of black material and spouting jets of vaporised ice.

As the comet approaches the Sun, the gases evaporate under the Sun's heat and form its coma, which scatters sunlight. The solar wind, a spray of charged particles from the Sun, blows this material into a tail, reaching typical lengths of 10 million kilometres, and in rare cases, several times that distance. The tail is always away from the Sun. Comets become visible only when they are close to the Sun – between two and three times as far away from the Sun as Earth. That's why astronomers can't be sure when the next new comet will come.

Whipple, often referred to affectionately as 'Dr Comet',

not only coined the evocative phrase 'dirty snowball', he also came up with an equally evocative idea: a comet is like a jet engine. Like the heated gases erupting from a jet engine, the evaporating gases from the nucleus exert a force on the nucleus. This force gives the comet its independent thrust. 'When I first realised the jet action of comets', 79-year-old Whipple told *Time* magazine in 1985, 'Boy! That was a thrill.'

The most famous of all comets is Halley's comet. It's named after the 18th-century British astronomer Edmond Halley, who first calculated its period and successfully predicted its return in 1758. It travels in a giant orbit that takes about 76 years. It was last seen in 1986, and is expected to return in 2061. It's a long wait for Halley's, but we never know – we may be lucky enough to see one night a brilliant comet arcing through the sky trailing a magnificent tail. It will outshine everything in the sky. When Halley's comet was here in 1910, Earth passed through its tail without any effect. It is a different matter if a comet head hits the Earth.

Comets usually pass Earth with speeds greater than 160,000 kilometres per hour. If a comet a couple of kilometres across hits Earth with such a speed, it will gouge out a hole as big as a large city, spewing out so much dust in the atmosphere that the Sun will be blocked out for months. If it hits an ocean, a tidal wave up to 1 kilometre high and travelling at many hundreds of kilometres per hour will submerge most low-lying regions. But the worst threat is fire – fire caused by debris thrown into the atmosphere when the comet head explodes before hitting Earth. Burning forests and cities will throw soot into an already clogged atmosphere. Then there will be an acid rain – a rain of toxic gases and metals. Most plants will die, followed by marine creatures that live near the surface, nearly ending life on the planet.

The comet did it

Back to Tunguska. The scary scenarios of a comet striking Earth are hypothetical, but the devastated Siberian taiga is a reality. The Soviet Academy of Sciences was determined to find an answer, and in 1954 sent Kirill P. Florenskiy, a geochemist, to survey the Tunguska site. Using modern cartographic equipment, Florenskiy made an aerial survey of the region and drew up a new, more accurate map. In 1957 the Russian mineralogist A.A. Yavnel microscopically analysed soil samples brought back by Kulik in 1929 and 1930. In some samples he discovered magnetite globules 30 to 60 micrometres in size. The samples also showed a small percentage of nickel and traces of cobalt. The results suggested that an iron meteorite had fallen in Tunguska.

These results encouraged the Academy to organise a new expedition. In the summer of 1958 Florenskiy headed the fifth Tunguska expedition, the first after the Second World War. The main aim was not to look for visible meteorite fragments but to collect micro-meteorites in the soil. The expedition failed to confirm the existence of magnetite and nickel globules detected by Yavnel. It was discovered later that Kulik's samples had been contaminated with samples from the iron meteorite that fell in the Sikhote-Alin mountains in

Siberia on 12 February 1947. The soil samples from this iron meteorite – which destroyed an area of more than a square kilometre and created many craters, the largest being 26.5 metres in diameter and 6 metres deep – had been stored with Kulik's samples. In his report to the Academy, Florenskiy rejected Kulik's hypothesis that the explosion took place on the ground, and suggested that it had taken place in the air some height above.

Florenskiy headed another expedition to Tunguska in 1961–62, which analysed in detail the pattern of devastation of the forest and established the trajectory of the fireball. However, by 1960 Florenskiy was convinced that Tunguska had been flattened by a comet. 'Many facts favor the view that the colliding body was a comet: the unusually loose structure, which led to breakup in the atmosphere; the dust tail, pointing away from the sun, which caused unusual sunsets over nearly all of Europe; the nature of the orbit; and lack of big fragments', he wrote in *Sky & Telescope* magazine. He pinpointed the location of the explosion at an altitude of 5 kilometres, southeast of the centre of destruction on the ground, and suggested that the collision was not head-on, and nor did the comet directly follow Earth, for it struck Earth almost squarely on the side.

He explained the explosion in Earth's atmosphere by saying that the sudden stopping of a body moving at 18,000 kilometres per hour releases enough heat to vaporise it instantaneously. If the body is loosely compacted and contains volatile matter such as gases or ice, deceleration in the air may cause explosive vaporisation without the body striking the ground, he suggested.

Florenskiy established that after the explosion the Tunguska forest burned for at least five days. A weak southeasterly wind at 7.2 to 18 kilometres per hour was blowing on that day, but the fire was preceded by a dry spell and therefore spread rapidly through the treetops. The fire died out because of the unfavourable weather conditions.

'We found that branches which had been not thicker than a fountain pen before the catastrophe still retained marks of their injuries', he said. 'The damage is noticeable on the upper portions of the branches, thus making it impossible to associate these injuries with ordinary fires. The severity of these injuries diminished significantly with increasing distance from the epi-centre.' He estimated that the energy required for the observed injuries would be 5 to 15 calories per square centimetre. This value could not be significantly higher, since this would lead to marked charring of the bark, and no such phenomenon was observed. 'Approximately the same energy is required to ignite dry forest debris, and this could lead to forest fire', he noted. Studies of nuclear explosions have shown that a 10-megaton explosion would produce charring similar to the Tunguska forest in fir, pine and maple bark.

In the early 1960s Vassilii Fesenkov of the Soviet Academy of Sciences' meteorite committee expanded the comet theory. He gave various reasons in support of his belief. These can roughly be grouped into four categories:

First, craters found by Kulik are now not considered to be places of the fall of fragments of the meteorite.

Despite extensive searches, no primary meteorite fragments have been found. It's easier to visualise a 'dirty iceberg' exploding to nothing than a rocky meteorite. Many magnetite and silicate globules (5 to 450 micrometres in diameter) found in the area were 'clearly of secondary origin'. They were most likely formed in the atmosphere due to rapid condensation of molten 'rain drops' as they drifted to the ground. The height of the comet's explosion – 5 to 6 kilometres above the Earth's surface – has been well established by the measurement of shock waves received at observatories in Irkutsk (Siberia) and Potsdam (Germany) and the six microbarograph stations in England.

Second, according to all evidence, this meteorite moved around the Sun in a retrograde direction – it was moving from south to north at a time when Earth was moving generally from north to south – which is impossible for typical meteorites. Meteorites rarely hit Earth in the morning, because the morning side faces forward in the planet's orbit. Usually the meteorite overtakes Earth from behind, on the evening side. Comets, on the other hand, have a wide range of orbits and speeds and could collide with Earth on the morning side, hitting head-on with a speed of about 145,000 kilometres per hour. The Sun's glare prevented any sighting of the comet before it hit Earth, because its direction and the angle of strike towards Earth were from behind the Sun.

Third, the most probable explanation for the brilliant night skies observed in the British Isles, Europe and Western Russia would be that the meteorite was actually a little comet with a dust tail pointing away from the Sun. The dissipation of this tail in the atmosphere greatly increased the night sky brightness. The dust particles causing the night glows were at a height of several hundred kilometres above the Earth's surface and did not behave as ordinary meteor showers.

Fourth, a marked decrease in the air's transparency, recorded two weeks after the explosion, was caused by the loss of several million tonnes of dust from the comet during its flight through the atmosphere.

Fesenkov told *The New York Times* on 20 November 1960 that 'the recent study suggests that explosions took place over at least three points in the area'. He said that in one of the Soviet studies, an experimental model was used to simulate the forest and miniature explosions were set off in the air. These experiments suggested that the head of the comet, a 'small one', consisted of dust and frozen gas in one or more extremely dense clouds several kilometres in diameter. 'The total weight is thought to have been more than 1,000,000 tons', *The Times* reported. 'When it hit the atmosphere the resulting explosion is thought to have been comparable in force to that of an equal amount of TNT.' It's worth noting that Fesenkov's estimate of the blast's energy – 1 megaton – is similar to the estimate made by Astapowitsch in 1933.

Further support for the comet theory came in 1975 from Ari Ben-Menahem, an Israeli scientist, who re-analysed the old seismographs of the Tunguska explosion and compared them with a series of air explosions from nuclear tests at the USSR test site at Novaya Zemlya. He concluded that the explosion took place 8.5 kilometres above the ground and had energy of about 12.5 megatons.

A year later, David Hughes, a British astronomer, estimated the comet's nucleus diameter at about 40 metres, much smaller than the diameter estimated for visual comets. The small diameter explained why the comet wasn't seen as it approached on its collision course to Earth. 'A cometary nucleus of this size will hit Earth about every 2,000 years, the rarity of the event giving ample justification for visiting Tunguska again', he said.

In a further study, Hughes and his colleague John Brown noted that, although the temperature produced by the burn-up of the comet in the atmosphere would have been no more than a few million degrees, too low for nuclear reactions, this temperature was high enough to produce nuclear-like effects such as the production of X-rays, gamma rays and highly accelerated charged particles. Even if the Tunguska body did cause nuclear effects, it does not mean it was not a comet. 'The Tunguska explosion was by an impacting small comet and [...] nothing more exotic needs to be invoked', Hughes and Brown concluded.

After a more recent analysis of the unusual sky glows, Vitalii Bronshten of the Committee of Meteorites of the Russian Academy of Sciences concluded that 'the cause of the glow was secondary scattering of sunlight by the dust constituting the envelope of the Tunguska comet'. Bronshten, one of the main contemporary supporters of the comet theory, first calculated the volume density of the dust ejected by the nucleus of a comet like Halley's at a distance of several thousand kilometres, and then its transfer to the west in Earth's gravitational field, taking into account deceleration caused by the atmosphere. He showed that the bigger dust would reach the British Isles in six hours. Smaller particles could cover greater distances but they could not produce a noticeable scattering of light. This is the reason why the glowing area had a western border: night glows were limited to the British Isles, Europe and Western Russia.

If the Tunguska object was indeed a comet, then it must be a comet known to astronomers. In 1978 the Slovak astronomer Lubar Krésak suggested that a piece of comet Encke had exploded at Tunguska. He based this idea on the fact that the fireball exploded at the peak of one of the most intense annual daytime meteor showers in late June, which has long been thought to derive from this comet. Encke's comet is named after the mathematician who investigated its orbit. The German astronomer Johann Encke was born five years after the comet was discovered in 1786, but he showed in 1822 that it has a period of three years and four months, the shortest known.

Was it a 'fair dinkum' comet?

In 1975 Fred Whipple questioned the possibility of a comet striking Tunguska. He estimated that if we take

75

the mass of the comet as 1 million tonnes, as suggested by Fesenkov, the chance of such a comet striking Earth every 100 years would be about 1 in 20,000. 'It appears unlikely, therefore, that the Tunguska explosion was produced by a *bona fide* active comet a hundred or so meters in dimension ... more likely, however, the Tunguska object was an inactive, low-density, friable body ... There is no reason to suspect it was interstellar.'

Most contemporary scientists also reject the idea of a comet. 'Comets are fluffy in comparison with asteroids and burn up quickly in the atmosphere', Richard Stone writes in *Discover* magazine. 'For one to have produced an explosion as big as the one over Tunguska, it would have started out as a million-tonne object. The vast swath of gas and dust left by such an object on its way down might well have shut out the sun or altered the climate.' Zdenek Sekanina, an expert on comets at NASA's Jet Propulsion Laboratory, agrees: 'The effect on life on Earth would have been horrendous. It would have been a global catastrophe, comparable to nuclear winter. The effects on mankind would have been so overwhelming that we could not discuss the topic, because we would not be here.' You and I are still discussing the topic; therefore, the Tunguska fireball was not a comet. QED.

Although the probability of such low-density objects colliding with Earth is obviously quite small, there is still hope for the comet theory – and it comes from Down Under. Assuming a speed of 108,000 kilometres per hour, as the Australian scientists Duncan Steel and Richard Ferguson present their case, seven hours before the impact the Tunguska object would have been about

750,000 kilometres from Earth. Active comets produce tails that stretch millions of kilometres away from the Sun, so it is possible that there could have been an encounter between Earth and the tail of the comet. This encounter could produce an aurora in the hours before the impact. But did anyone observe an aurora seven hours before the Tunguska blast?

Steel, a well-known authority on the threat posed by asteroids, was at a conference on asteroids, comets and meteors in Sweden in 1989, when the Russian scientists Nikolai Vasiliyv and G. Andreev circulated a short report on Soviet research on Tunguska. He was intrigued by the following paragraph in the report:

> A special item in this respect could be the search for the original diary entries by Mouson who observed auroras from near the Erebus volcano at Antarctica during the summer of 1908. There is information in Shackleton's accounts that on June 30, Mouson registered an aurora which he visually considered to be anomalous. Unfortunately, Shackleton's accounts do not contain further details.

Steel soon figured out that 'Mouson' was 'Mawson', after transliteration from the Latin script to Cyrillic and then back again, and 'the summer of 1908' was in fact 'the Antarctic winter of 1908'. Coincidentally, at that time Steel was at the University of Adelaide, where Mawson's notebooks from the Antarctic Expedition of 1907–09, led by Sir Ernest Shackleton, are archived in the Mawson Institute for Antarctic Research. Mawson (later Sir

Douglas Mawson) was a young geologist on the expedition, and kept a diary of his observations. Steel and Ferguson made extensive searches of Mawson's diaries and all other expedition papers, but failed to find any record of aurora australis at the time of the Tunguska blast. However, they found a record of an exceptional aurora seven hours before the blast. Was this aurora caused by the Tunguska fireball? If yes, then the fireball was, as Mawson would have said, a fair dinkum (genuine) comet.

CHAPTER FOUR

ASTEROIDS
BEHAVING BADLY

Once called boring 'vermin of the skies', asteroids are now star attractions for astronomers around the world. This attention is worthy of their name, which is Greek for 'starlike', but it has more to do with their sheer number and their destructive power than their cosmic beauty. These pockmarked giant peanut-like rocks are in fact leftovers from the formation of the planets.

The largest three asteroids are Ceres, Pallas and Vesta, with average diameters 930, 520 and 500 kilometres respectively. About 200 asteroids are larger than 100 kilometres across; 800 larger than 30 kilometres. About a million are 1 kilometre or more in diameter; and billions are of boulder or pebble size. Most of the asteroids orbit within a vast, doughnut-shaped ring between Mars and Jupiter, known as the main belt.

Occasionally, a collision may kick an asteroid out of the belt, sending it onto a dangerous path that crosses Earth's orbit. These stray asteroids take up an orbit that loops past Earth, and are called 'Earth-crossers'. This knowledge frightens astronomers. What if one of them comes too close to Earth? What cataclysm would such a rogue rock cause if it slams into Earth? The number of asteroids is very large, but the space they occupy is enormous. Most asteroids stay millions of kilometres apart. It's not like *Star Wars* or *Star Trek* spaceships weaving their way through flying rocks. But real collisions are possible with a spacecraft or Spaceship Earth.

There are believed to be about 1,800 Earth-crossers, or near-Earth asteroids, a kilometre or greater in width. Only 500 or so have been discovered so far, but astronomers hope to identify almost all of them by the end of this decade. The largest presently known is 1036 Ganymed, with a width of about 41 kilometres. There may be as many as a million near-Earth asteroids 50 metres and larger. The chances of these rocks hitting Earth are small, but even one of the smaller asteroids could destroy a large city.

Don't panic – none of them is on a collision course yet. On 14 June 2002 a football-field-sized asteroid came within 120,000 kilometres of Earth. It was the biggest asteroid in decades to get this close to us. For you and me, 120,000 kilometres is, well, a long way away. For astronomers who coin words like 'Earth-crosser' it is a hair's-breadth; for us, a hair-raiser indeed. The errant asteroid was discovered three days after it sped by Earth at 38,000 kilometres per hour. If it had struck Earth, it would have caused a Tunguska-like explosion. Did such a strike indeed take place on 30 June 1908?

Pointing an accusing finger

After meticulously reviewing five decades of research on the Tunguska fireball, in 1983 Zdenek Sekanina put together a new analysis of the orbit, atmospheric entry and explosion of the interstellar body. He ruled out a comet as a suspect and pointed an accusing finger at an asteroid about 90 to 190 metres across. The asteroid came in from a direction close to 110 degrees east of north at an angle about 5 degrees above the horizon, and exploded at about 8 kilometres above the ground. Its speed when it entered the atmosphere was about 108,000 kilometres per hour.

In his 32-page seminal paper in *The Astronomical Journal*, Sekanina remarked that evidence on the forest devastation in the area of fall of the Tunguska fireball left no doubt that it exploded in mid-air and then completely disintegrated in the atmosphere. The evidence included the absence of impact craters and/or sizeable debris in the area of the fall, a near-perfect radial symmetry of the region of flattened trees to within at least 15 kilometres from its centre, and the presence of standing bare trunks (Kulik's 'telegraph poles') at the very centre of the radial of devastation.

Since the evidence pointed to only one enormous outburst, Sekanina concluded that the fireball did not fracture in flight. If such break-ups had in fact occurred, the resulting sequence of explosions would have diminished the enormous power of the final explosion.

The explosion released almost instantaneously energy in excess of 12 megatons. The eyewitnesses saw this energy, which was enough to wipe out even modern London or Tokyo, as a fireball 40 times brighter than the noonday sun. 'This conclusion is supported by accounts of a dazzling blaze of light in the sky, described in eyewitness reports from places near the centre of the explosion', Sekanina said.

The fireball dissipated a million tonnes of small particles in less than one-tenth of a second in Earth's atmosphere. Winds assisted by the expanding shock front dispersed the dust in the stratosphere, which caused the bright skies reported in the aftermath of the fall.

Sekanina rejected the notion that the Tunguska fireball was a comet. Because of the high speed at which the object was travelling when it entered the atmosphere, it resisted a very high air pressure before exploding. It is inconceivable that a comet, known for its extreme fragility, could have survived such a high pressure. The fireball, therefore, must have been a denser, stony object to survive its journey to the Siberian sky. It was probably a small member of the Apollo asteroids, believed to be nuclei of comets that have lost their volatile components.

Critics of the asteroid theory said that Sekanina's analysis was mainly based on eyewitness accounts, most of which were recorded at least two decades after the event, and seismic records. 'You can't make a sophisticated model from poor data', the American meteor specialist Richard McCrosky told *Sky & Telescope*. 'It seems everything he assumes must be true for his conclusion to be right.'

In the same year the American atmospheric scientist Richard Turco suggested that the bright nights were caused by noctilucent clouds, silvery clouds at high altitude that shine at night. On 30 June 1908 the wind was blowing in the right direction for the dust associated with the fireball to reach western Europe. The dust would then have settled in the atmosphere at an altitude of about 80 kilometres and remained there for several days. These dust particles and the water vapour that was also deposited by the fireball contributed to enhance cloudiness. A stony asteroid would have been too dry to provide the water needed for such clouds. Turco's analysis favoured an icy comet nucleus.

More recently, Vitalii Bronshten rejected the asteroid theory on the ground that an analogy with nuclear explosions suggested that, even upon a very strong explosion, a stony asteroid should break up into fragments of various sizes. These fragments could not completely vaporise in the intense heat of the explosion or as a result of their fall to the Earth. Some of these fragments should survive, but even after many careful searches no such fragments have ever been found. Bronshten also suggested that the noctilucent clouds, which he agreed were observed, could never gain such brightness to result in light nights.

An asteroid masquerading as a planet

My Very Educated Mother Just Served Us Nine Pizzas – the mnemonic you learned at school to list the order of the nine planets (outwards from the Sun) may soon be without Pizza for Pluto. Your very angry mother would be serving nine nothings.

Pluto isn't rocky like Mercury, Venus, Earth and Mars, nor is it a gas giant like Jupiter, Saturn, Uranus and Neptune. It is a relatively tiny ball of ice; with a diameter of 2,274 kilometres it is smaller than the Moon. Pluto's orbit is also peculiar. The other outer planets orbit the Sun in roughly circular orbits, but Pluto's orbit is elliptical, which at times brings it closer to the Sun than Neptune. The oddball Pluto is not really a planet – that's what some astronomers have been arguing for years. They say it's the largest object in the Kuiper belt, the outpost of icy asteroids beyond the orbit of Neptune.

The existence of such a belt was first suggested in the early 1950s but the first Kuiper belt object, called 1992 QB1, was found in 1992. This small icy body, similar in size to an asteroid, suggested that there might be more than just Pluto in the distant reaches of the solar system. Since then hundreds of objects like QB1 have been found in the Kuiper belt. Their diameters range from 50 to almost 1,200 kilometres. Though these objects are smaller than Pluto, they are thought to be similar to Pluto in composition.

The Kuiper belt is believed to contain about 100,000 objects larger than 100 kilometres across. This swarm of Pluto-like objects gives weight to some astronomers' argument that Pluto is not a planet but a very large ball of ice. Even those astronomers who defend Pluto's status as a planet agree that were it discovered today it probably wouldn't be called a planet.

The discovery in 2004 of the most distant planet-like object in the solar system made the Pluto debate more interesting. Named Sedna after the Inuit ocean goddess, the new object is three-quarters the size of Pluto and way outside the Kuiper belt. The discovery will now hot up the debate about whether Pluto is the puniest planet or the king of the Kuiper belt. This debate has a historical precedent. More than two centuries ago when Ceres was discovered, it was also proclaimed a planet.

Piazzi's 'planet'

The discoverer was defiant: 'I have the full right to name it in the most convenient way to me, like something I own. I will always use the name Ceres Ferdinandea, nor by giving it another name will I suffer to be reproached for ingratitude towards Sicily and its king.'

Shortly after nightfall on 1 January 1801, Giuseppe Piazzi, a monk and director of the brand new Palermo Observatory built atop a 12th-century tower in the royal palace of Sicily, pointed his shining brass telescope at stars in the constellation Taurus and observed an unfamiliar, faint, starlike object. Later observations and calculations showed it to be the 'missing planet' between Mars and Jupiter. He named it Ceres Ferdinandea – Ceres for the patron goddess of Sicily, and Ferdinandea for his royal patron, King Ferdinand of Naples and Sicily.

The discovery caused a sensation in Europe. The burning question was: what should the new planet be named? Napoleon even discussed it with the celebrated French mathematician and astronomer Pierre Laplace. Some French astronomers suggested Piazzi, while some Germans favoured Juno or Hera. But Piazzi was adamant that he had the right to name it.

The story of Piazzi's discovery starts in 1772, when Johann Titius, a professor at Wittenberg in Germany, discovered a remarkable numerical relationship between the distances of the planets from the Sun. He pointed out that the numbers in the series 0, 3, 6, 12, 24, 48, 96, when added to 4 and divided by 10, produced the series 0.4, 0.7, 1, 1.6, 2.8, 5.2, 10. If Earth's distance from the Sun is set at 1 astronomical unit (about 150 million kilometres), then these numbers give the distances of the six planets known at the time, except for position 2.8. Titius suggested that this gap belonged to still undiscovered satellites of Mars.

That same year, the German astronomer Johann Bode picked up Titius's rule and quoted it without any acknowledgement in his astronomy textbook. However, he suggested a new planet for the gap at 2.8. The rule is now known as Bode's law. Although Bode also carried out other astronomical investigations, he is remembered today for popularising a relationship that he did not originate.

When the celebrated German-British astronomer William Herschel discovered the planet Uranus in 1781, it also fitted Bode's law (continuing the Titius series by doubling 96 for Saturn, that is, 192; when added to 4 and divided by 10, this gives 19.6, which is close enough to 19.2, the actual distance of Uranus from the Sun in astronomical units). Astronomers now strongly felt that another planet was to be found between Mars and Jupiter.

The Hungarian astronomer Franz von Zach so strongly believed in the 'missing planet' that he tried to calculate its orbit by using Kepler's laws, but one element that might reveal its location – the longitude – eluded him. In 1785 he wrote to Bode: 'I am having much the same success as the alchemists in their search for gold; they had everything except the vital factor. Thus I too believe that I am in possession of all the elements of the orbit of this still unknown planet, except one; this alone now keeps me amused, and although one may not find gold in one's wanderings, one does occasionally come across a chemical process.'

In 1787 von Zach took a solo search for the planet, but without success. 'It cannot be a matter for one or two astronomers to scrutinise the entire Zodiac', he wrote in *Monatliche Correspondenz* (Monthly Correspondence), the world's first astronomical journal, which he founded. The hunt for the missing planet began in earnest when, in 1800, von Zach organised a group of 24 astronomers who called themselves the 'celestial police'. They divided the entire Zodiac into 24 zones. The zones were then allocated to the members by lot. Each member was to be responsible for drawing up a star chart for his zone.

'Through such a strictly organised policing of the heavens, divided into twenty-four sections, we hoped eventually to track down this planet, which had so long escaped our scrutiny – supposing, that is, that it existed and could be seen', von Zach wrote in *Correspondenz*. Before 'such a strictly organised policing' of the heavens could get under way, surprising news arrived from Palermo on the island of Sicily.

Giuseppe Piazzi, a Theatine monk who entered the order in 1764 at the age of eighteen, received his early training in philosophy but later in life took up mathematics and astronomy. In 1780 he was called to the chair of higher mathematics at the Academy of Palermo. There he soon obtained a grant for an observatory, and he went to England in 1788 to buy instruments for it. He commissioned Jesse Ramsden, the greatest of the instrument-makers, to make him a 1.5-metre vertical circle of unique design for measuring the altitudes and azimuths of stars by micrometer microscopes. While in England he became acquainted with Herschel and had the unique 'privilege' of falling off the high wooden ladder at Herschel's large reflecting telescope and breaking his arm.

Once the Ramsden circle, the masterpiece of 18th-century technology, was installed at Palermo in 1791, Piazzi started his painstaking work of cataloguing stars. In 1803 he published his first catalogue containing 6,784 stars, and in 1814 a second one containing 7,646 stars. However, Piazzi's major accomplishment did not involve the stars at all. It was the discovery of the 'missing planet'.

New century's gift

On 1 January 1801, the first evening of the new century, Piazzi observed an unfamiliar point of light in the sky. He thought that the object might be a new star. Over the next three evenings he observed it again and noticed that it had shifted its position at the same rate as on the preceding days. Piazzi was now sure that it was not a fixed star. Thinking it might be a comet, he continued to follow it until 11 February when an illness cut short his work. However, on 24 January he announced his discovery to Bode, the French astronomer Joseph Lalande, and his friend Barnaba Oriani, director of the Brera Observatory in Milan.

He confided only to Oriani that it might possibly be a new planet: 'I have announced this star as a comet, but since it is not accompanied by any nebulosity and, further, since its movement is so slow and rather uniform, it has occurred to me several times that it might be something better than a comet. But I have been careful not to advance this supposition to the public.'

Oriani replied: 'I congratulate you on your splendid discovery of this new star. I do not think that others have noticed it, and because of its smallness, it is unlikely that many astronomers will see it.' Bode thought that Piazzi's discovery marvellously fulfilled his prediction of a planet between Mars and Jupiter. Von Zach was elated and reported the news in *Monatliche Correspondenz*, under the heading 'On a long supposed, now probably discovered, new major planet of our solar system between Mars and Jupiter'.

But Bode and von Zach could not verify the discovery. Such was the state of the postal service in those days that Bode did not receive Piazzi's letter until 20 March. By now the new planet had ceased its retrograde motion and had begun to advance, and had moved near enough to the Sun that it could not be seen. Everyone eagerly awaited its emergence from the other side of the Sun in July. Herschel was the first to search for it in July, and he and many other astronomers continued their search for months. The planet was missing again.

It required the genius of the German mathematician Carl Friedrich Gauss to recover Piazzi's lost planet. Gauss, then 24 and at the beginning of a brilliant career that placed him in the company of Archimedes and

Newton, calculated the orbit of the planet from Piazzi's few observations. Gauss's calculation of the planet's position was so accurate that on 31 December 1801, within a few hours of one another, von Zach and the German amateur astronomer Heinrich Olbers independently recovered the lost planet.

Piazzi named it Ceres Ferdinandea, but it was soon shortened to Ceres. King Ferdinand wanted to strike a gold medal with Piazzi's effigy, but the astronomer requested the privilege of using the money to buy a much-needed equatorial telescope for his observatory.

The discovery of Ceres posed a problem for astronomers. Herschel's observations showed that it was a most unusual planet, being too small to show a planetary disc. The problem was compounded when, on 28 March 1802, Olbers discovered 'another Ceres', a body that was also orbiting in the 'gap' between Mars and Jupiter. Olbers, a medical doctor by profession, named it Pallas. The discovery was most perplexing.

Olbers wrote to Herschel: 'Could it be that Ceres and Pallas are just a pair of fragments, or portions of a once greater planet which at one time occupied its proper place between Mars and Jupiter, and was in size more analogous to the other planets, and perhaps millions of years ago, either through the impact of a comet, or from an internal explosion, burst into pieces.'

Within a month of the discovery of Pallas, Gauss calculated its mean distance from the Sun, almost the same as Ceres. He also noted that Ceres and Pallas had many characteristics that made them quite unlike planets. Bode was not yet convinced that his law (which he had unceremoniously pinched from Titius) was not true. He wrote to Herschel: 'I hold myself convinced that Ceres is the eighth primary planet of our solar system and that Pallas is a special exceptional planet – or comet – in her neighbourhood, circulating around the sun. So there would be two planets between Mars and Jupiter, wherever since 1772, I have expected only one; and the well-known progressive order of the distances of the planets from the sun, is by this fully proved.' Herschel was not influenced by Bode's plea for his law. He was now convinced that Ceres and Pallas represented a new and different class of celestial bodies.

He also believed that since Ceres and Pallas did not occupy the space between Mars and Jupiter with 'significant dignity', these new bodies were not worthy of the name 'planet'. He proposed that they should be given the name 'asteroid' (from the Greek *asteroeides*, 'starlike'), since they are intermingled with, and similar to, the small fixed stars. He went on to advocate three forms of celestial bodies – planets, asteroids and comets.

Most astronomers now accepted that Ceres and Pallas were not planets, but Piazzi was not happy with Herschel's celestial hierarchy of planets, asteroids and comets, and retorted: 'Soon we shall be seeing counts, dukes and marquesses in the sky.' He suggested the name 'planetoids', pointing out that 'asteroid' would be more appropriate for 'little stars'. The term 'asteroid' has persisted, but they are sometimes also referred to as 'planetoids' or 'minor planets'.

Von Zach's 'celestial police' did not give up their quest, and continued to scan the heavens through their telescopes. Their efforts were rewarded when German astronomer Karl Harding discovered the third asteroid, Juno, on 1 September 1804. Olbers added the fourth, Vesta, on 29 March 1807. Both were too tiny to be qualified as a planet. The talented Olbers also suggested an idea which has passed the test of time: noting that the brightness of Ceres and Pallas appears to vary from one observation to another, he said that asteroids have an irregular rather than round shape. To Olbers, asteroids appeared like rocks tumbling through space. And he was right. However, he is now best known for the Olbers paradox, the answer to the deceptively simple question: why is the sky dark at night?

The thousandth asteroid was discovered in 1923 and was named Piazzia in honour of Piazzi (who had died nearly a century earlier in 1826 at the age of 80). Since that time, no year has passed without the discovery of new asteroids. Now discoverers do not have to fight for their right to name their discovery. Once the precise orbit of an asteroid is determined, it is given a permanent catalogue identification consisting of a number that denotes its order of entry, which is usually followed by a name proposed by the discoverer, for example, 1 Ceres, 2 Pallas. Until the discovery is approved by the International Astronomical Union, the asteroid is provisionally known by its year of discovery followed by two letters, and numbers if necessary, indicating the date and sequence of discovery; for example, 1950 DA (the letter 'D' signifies that it was discovered in the period 16–28/29 February, and the letter 'A' that it was the first discovery during that period).

Target Tunguska

In 1993 NASA scientists Christopher Chyba and Kevin Zahnle and their colleague Paul Thomas from the University of Wisconsin gave the asteroid theory for the Tunguska explosion new weight and rigour.

Their computer simulations to explain the pattern of trees blown down showed that the explosion released about 15 megatons of energy in the atmosphere at an altitude of about 8 kilometres, but did not crater the Earth's surface. They then examined the entry of three classes of asteroids (stony, iron and carbonaceous) and two classes of comets (short-period and long-period) starting with 15 megatons of kinetic energy. Their simulation showed that cometary nuclei and carbonaceous asteroids explode far too high in the atmosphere to account for the blast, and iron asteroids hardly fragment and so hit the ground at high speed. Only the stony asteroids would create a Tunguska-like explosion.

In their innovative analysis, the researchers included the effects of aerodynamic forces on an asteroid: as the asteroid moves deeper into the atmosphere, atmospheric drag on it increases. When the drag exceeds the asteroid's strength, the asteroid crumbles and begins to flatten like a pancake. The increasing surface of the fragmented asteroid experiences a sharper rise in drag. Increasing drag slows down the asteroid, which in turn spreads it even more. At the same time, atmospheric density rises with decreasing altitude, creating more drag. These increasing forces stop the asteroid abruptly in the atmosphere. The asteroid explodes like a bomb, and within a fraction of a second megatons of kinetic energy is converted into high temperatures and high pressure. The asteroid vaporises. A shock wave races outwards.

Taking account of these effects, the researchers calculated that a stony asteroid about 30 metres in diameter and moving at 54,000 kilometres per hour would explode at a height of about 8 kilometres, the same height at which the Tunguska body apparently exploded. A smaller one would have exploded much higher, and a larger one would have created an impact crater.

The weak, fast-moving, easily crumbled comets do not penetrate the atmosphere deeply and are unlikely to approach the altitude of the Tunguska explosion. 'Even if the comets had strengths comparable to stony asteroids, they still could not fit the Tunguska observation', the researchers maintained. For example, if a 15-megaton incident comet is assigned an anomalously low asteroid-like speed of 54,000 kilometres per hour, it would completely exhaust itself before it reached an altitude of about 16 kilometres. It would have also caused far less surface destruction.

The measurements of Halley's comet by the *Giotto* spacecraft in 1986 showed that it had a density of between 0.6 and 1 gram per cubic centimetre. In their simulation the Chyba trio used a density of 1, but said that values as low as 0.3 might be possible. Such lower-density objects would airburst even higher. 'Tunguska was probably a fairly strong, dense, asteroid-like object, but probably not as strong or dense as iron', the researchers concluded in their report in *Nature*. 'Carbonaceous asteroids and especially comets are unlikely candidates for the Tunguska object.' However, their simulation did not completely rule out an unusually fast iron asteroid or a very strong carbonaceous asteroid.

Chyba's team also supported Turco's 1983 observation that bright nights were caused by noctilucent clouds. They said that the air heated by the explosion injected enough water into the stratosphere for noctilucent clouds to be produced.

Further support for the asteroid theory came from Jack Hills and Patrick Goda of the Los Alamos National Laboratory. In their general study of fragmentation of asteroids, they found that a stony asteroid must be greater than 200 metres across, much larger than the Tunguska object, to hit the ground. They agreed that the Tunguska evidence ruled out a comet, and that furthermore there is no question but that the object was a stony asteroid and not an iron asteroid. Comets start fires more easily than asteroids, but the Tunguska asteroid generated enough heat to ignite pine forests. 'However, the blast wave from an impacting asteroid goes beyond the radius in which the fire starts', they said. 'The blast wave tends to blow out the fire, so it is likely that the impact will char the forest (as in Tunguska), but the impact will not produce a sustained fire.'

Henry J. Melosh of the University of Arizona in Tucson, commenting on the Chyba team's research, said that they had 'wrapped up' the most believable explanation for the Tunguska event. 'Substantial progress has thus been made in reducing the Tunguska explosion from the realm of the near-miraculous to a natural, although rare, occurrence', he said.

Chyba declared in *Astronomy* in December 1993: 'According to this picture, Tunguska goes from an exotic event demanding the invocation of UFOs or black holes to a completely normal, predictable outcome for a stony asteroid entering the atmosphere with a typical velocity. This understanding is important, for it allows us to assess the contemporary hazard posed by small asteroids and comets colliding with Earth.'

Fifteen years after proposing his original asteroid theory, in 1998 Zdenek Sekanina revisited his earlier analysis and concluded that the interpretation of the Tunguska event as a fall of a small asteroid is 'not only plausible, but virtually certain'. However, he said that one issue where he was not prepared to take sides was whether the object was a stony or a carbonaceous asteroid.

In his new analysis, Sekanina noted that the collision of comet Shoemaker-Levy 9 with Jupiter in 1994 showed that the mass of a comet that enters the atmosphere of a planet, such as Earth or Jupiter, apparently needs to be more than 100 million tonnes in order to trigger a powerful explosion at the end of its journey. Smaller comets are likely to dissipate their mass in atmospheric flight and would end up with no appreciable mass at low altitudes. By contrast, the study of a number of fireballs pointed out that initial masses as small as 10 kilograms exhibited terminal flares. Various studies of the inter-relation between a fireball's altitude, its pre-explosion speed, and the aerodynamic pressure at which it explodes corroborate the Tunguska fireball's pre-explosion mass of 1 million tonnes.

The question of a cometary hypothesis, Sekanina

implied, 'now becomes mute'. He also noted that the Tunguska fireball was dwarfed by the comet Shoemaker-Levy 9 when measured by the amount of released energy. 'On the other hand, the Tunguska event directly involves the issue of a threat to our civilization', he said.

Sekanina concluded his analysis by summarising eight points that made his asteroid theory broad-based and not based merely on eyewitness reports, as claimed by his critics: (1) its mid-air explosion is similar to the terminal flares of fireballs observed photographically; (2) the pressure at the point of explosion, estimated at about 200 times normal atmospheric pressure, is consistent with a value that is expected from similar fireballs; if the object were a comet, the pressure would be about 2,000 times atmospheric pressure, entirely out of a plausible range for a fragile comet; (3) the pre-explosion velocity of 36,000 kilometres per hour is the same as determined by seismic observations and a laboratory simulation of the uprooted forest; (4) this velocity range rules out a comet-like orbit; (5) existing limited evidence on the event is inconsistent with a fragmentation pattern typical for comets; (6) a comet of such a magnitude would have been extremely rare, perhaps 10 to 100 times more so than an asteroid of the same explosion energy; (7) the limited evidence on the object's orbit is consistent with the orbits of the Earth-crossing asteroids, but not with the orbits of short-period comets; and (8) this orbital information is particularly unfavourable to the hypothesis that associates the object with Encke's comet.

In 2001 a team of Italian scientists looked at the Tunguska object from a different angle, based on an idea of the late Paolo Farinella (1953–2000). Using data from detailed analysis of all the available scientific literature, including unpublished eye-witness accounts that have never been translated from the Russian, and the survey of directions of more than 60,000 flattened trees, the Italian scientists plotted a series of possible orbits for the object. Of the 886 valid orbits that they calculated, 83 per cent of them were asteroid orbits, with only 17 per cent being orbits that are associated with comets.

This overwhelming data showed that the Tunguska object was indeed an asteroid. But if it was an asteroid why did it break up completely? According to one of the team, Luigi Foschini of the University of Trieste: 'Possibly because the object was like asteroid Mathilde, which was photographed by the passing Near-Shoemaker space-probe in 1997. Mathilde is a rubble pile with a density very close to that of water. This would mean it could explode and fragment in the atmosphere with only the shock wave reaching the ground.'

As the 25th anniversary of his 1983 paper – and the 100th anniversary of the Tunguska event – approaches, Sekanina reflects upon his ground-breaking conclusion that the Tunguska object was an asteroid:

I have not changed my views on the subject in the least, still considering the Tunguska's asteroidal nature as virtually certain ... My involvement with the subject was in fact quite peculiar. I would not have begun my research on this subject in the early 1980s, if it were not for the fact that the papers published through the 1970s – all strongly pro-comet – had attracted my attention. I am a comet physicist and I felt that if it was a piece of a comet, one should be able to learn something on the properties of cometary nuclei from the event. Thus, I began to study the topic believing that this was indeed a comet: otherwise I would have never got involved and I surely cannot be accused of initiating my research on this subject with a bias against the Tunguska's cometary origin.

My conclusion was simply the result of my findings: the more work I did, the more obvious it became to me that this was not a comet (just take the huge dynamic pressures to withstand!). By the time I was convinced completely, I invested so much time that I felt it would be a shame to drop the subject, even though it became essentially irrelevant to my own scientific interests (I never found asteroids to be much fun). And this was the sole reason behind my writing the 1983 paper. The rest is history.

He adds: 'Of course, the cometary-origin hypothesis still has its old proponents such as Vitalii Bronshten, but no new admirers.' And, of course, the major contribution of Russian scientists in the development of the comet theory is well recorded in the history of the enigma known as Tunguska.

Target Earth

Are we going to be hit by a Tunguska-like asteroid again? Astronomers suggest that the average frequency of impacts of this size (average width 75 metres) is 1 in 1,000 years. These asteroids explode in the lower atmosphere but release enough destructive energy to wipe out a large city.

The rate of impact decreases with the increase in size of the asteroid. The average interval between impacts of giant asteroids (average width 16 kilometres) is 100 million years. An impact of this magnitude could destroy an entire continent and trigger mass extinction of advanced forms of life. Such an impact is believed to have wiped out the dinosaurs around 65 million years ago (see Chapter Ten).

Do we really want to win an asteroid lottery? The probability of randomly picking six numbers in a lottery of 45 numbers is 1 in 4 million. The probability increases to 1 in 14 million if you have to randomly pick six numbers from 49 numbers, and to 1 in 19 million if you have to pick from 51 numbers. The probability of an asteroid impact, small or big, is 1 in 20,000, the same probability as for a passenger aircraft crash. From these odds it appears that the proverbial man in the street, if he is not run over by the proverbial bus (probability 1 in 100), will witness an asteroid impact long before he wins the big lotto. Why bother to buy a lottery ticket today?

Should we dismiss these risk probabilities as lies, damned lies and statistics, or lose sleep over the asteroid threat? What do the experts say?

'We do not know whether a large dangerous asteroid with our name on it is destined to hit [this] century', says Britain's Astronomer Royal, Martin Rees. 'The risk isn't large enough to keep anyone awake at night, but it isn't completely negligible either.'

The American planetary scientist Tom Gehrels has a similar opinion: 'The chances of a celestial body colliding with Earth are small, but the consequences would be catastrophic.'

'Nobody believed Chicken Little when he said that the sky was falling. But occasionally the sky does fall, and with horrendous effects.' That's what Eugene Shoemaker, who was chiefly responsible for alerting the world to the dangers of asteroid and comet impacts, said in 1994.

'Doubters can ask the dinosaurs for their opinion', advised a *Scientific American* editorial in November 2003.

Besides asking dinosaurs' expert opinion, what else could scientists do to stop a rogue asteroid from crashing into Earth? Once scientists have discovered an asteroid they can calculate whether it is headed our way. The plan to destroy it will depend upon how far away it is. Here are some of the plans to save us from the scenarios dramatised in the movies *Armageddon* and *Deep Impact*:

Nudge 'em. If a threatening asteroid is spotted one year before the expected collision, it would be possible to nudge it with conventional-chemical fuel missiles. A small change in the asteroid's speed, on the order of 36 metres per hour, will deflect the asteroid 6,000 kilometres – the radius of the target Earth – in one year.

Push 'em. Instead of a gentle nudge, some scientists prefer a stronger push by a nuclear-powered spacecraft that expels jets of plasma. The unmanned 'space tug' would rendezvous with the killer asteroid, attach to its surface and slowly push it so that it misses Earth.

Crush 'em. This ingenious plan suggests crushing the asteroid by placing a three-dimensional lattice made of millions of small tungsten balls in the path of the speeding offender. The collision would create enough heat to turn the asteroid into small, harmless rocks. The lattice could be launched into space by a rocket.

Cover 'em. Another ingenious plan is to wrap the rogue asteroid with a shiny plastic sheet like a giant potato in aluminium foil by sending a solar sail spacecraft that collapses around the asteroid and shrink-wraps it. An asteroid radiates heat into space after the Sun warms its surface, which imparts a tiny momentum to the asteroid, slightly shifting its orbit. A shrink-wrap or a spray of white chalk or black carbon powder across an asteroid's surface would change its reflectivity and hence the heat transfer from its surface. This heat transfer would change the momentum, which could be enough to change its path. After gift-wrapping it would be years before the asteroid changed its course.

Nuke 'em. If there is not enough time to prepare, some scientists suggest destroying the asteroid with a nuclear bomb. The idea is that the nuclear blast would melt material off the asteroid's surface, giving it a kick in the opposite direction. This is not really a good way of killing an asteroid: if the bomb detonates too close to the asteroid, it may explode it, creating millions of tonnes of radioactive dust and rubble. If the rubble rains on Earth, we may have to join the dinosaurs anyway. Neutron bombs – the bombs that kill people but leave buildings undamaged – offer a better alternative. Once the high-energy neutrons hit the asteroid, they would heat its surface. The vaporised material would deflect the asteroid from the collision path.

Burn 'em. This plan involves placing a huge aluminium concave mirror fairly close to the asteroid. The mirror would focus a beam of light on a small spot on the asteroid. The heat would vaporise a small part of the asteroid which would shoot into space, pushing the asteroid in the opposite direction. An 800-metre wide mirror could deflect an asteroid 3 kilometres in diameter.

Dig 'em. This simple but technically difficult plan requires placing a robot on the surface of the asteroid. The robot would dig rocks from the asteroid and hurl them into space, causing it to accelerate slowly in the opposite direction.

Robert Gold of Johns Hopkins University, who considers an asteroid or comet impact 'the greatest natural threat to the long-term survivability of mankind', has proposed a comprehensive Earth defence system designed to discover, catalogue and calculate the orbits of near-Earth objects, and to deflect potential hazards. His three-part SHIELD system consists of Sentries, spacecraft designed to search for and locate threatening objects, Soldier spacecraft to deflect or disperse the object, and an Earth-based control system to oversee the network. Gold's Soldiers would use one or more of the deflection techniques discussed above. Gold believes that SHIELD can be implemented within the next 10 to 40 years.

In 2002 NASA's Jet Propulsion Laboratory (JPL)

established its SENTRY programme, a highly automated collision monitoring system that continually scans the most current asteroid catalogue for possibilities of impact with Earth over the next 100 years. Whenever a potential impact is detected, it is analysed and the results are immediately published on JPL's NEO (near-Earth objects) Programme website: http://neo.jpl.nasa.gov/risk/

This website also keeps track of the asteroid 1950 DA. As the name suggests, it was discovered in 1950. It was observed for seventeen days but then it faded from view for half a century. It was rediscovered on New Year's Eve 2000. There is a 1 in 300 long shot that the 1.1-kilometre-wide 1950 DA would hit Earth in March 2880. Don't forget to pass the JPL website's good news on to your next 35 generations: 'There is no reason for concern over 1950 DA. The most likely result will be that St Patrick's Day parades in 2880 will be a little more festive than usual as 1950 DA recedes into the distance, having passed Earth by.'

On the matter of the doomsday rocks, Tom Gehrels has the last word:

Comets and asteroids remind me of Shiva, the Hindu deity who destroys and re-creates. These celestial bodies allowed life to be born, but they also killed our predecessors, the dinosaurs. Now for the first time, Earth's inhabitants have acquired the ability to envision their own extinction – and the power to stop this cycle of destruction and creation.

CHAPTER FIVE

TRACELESS TUNGUSKA

Since Kulik's first expedition in 1927, there have been numerous Russian and international scientific expeditions to the Tunguska explosion site, but no impact craters or substantial meteorite remnants have yet been found. Besides a flattened forest, did the fireball leave other fingerprints?

Those who believe that the Tunguska object was either a comet or an asteroid also believe that the mass of the object vaporised into microscopic debris when it exploded at some height above the ground. Such a theory would also account for the absence of any impact craters. After the explosion, some of the microscopic debris drifted westward from Tunguska, but the rest condensed into microscopic globules that rained on to the Tunguska taiga. Many scientists have tried to interpret these fingerprints.

Early attempts

Scientists' attempts to track marks and traces of Tunguska began in 1957 when the Russian mineralogist A.A. Yavnel analysed soil samples collected by Kulik in 1929 from the Tunguska site, but these samples were later proved to be of terrestrial origin. Later, mineralogist

O.A. Kirova, who was a member of Kirill P. Florenskiy's 1958 expedition, recovered both magnetite globules and various forms of silicate globules from samples collected from the region of the fall. The magnetite globules were either shiny or dull and their sizes varied from 5 to 450 micrometres. The silicate globules varied in appearance from opaque to completely transparent and their sizes varied from 20 to 350 micrometres. The opaque silicate globules also contained large numbers of gas bubbles and traces of iron oxide. Both types of globule are characteristic of the particles produced by the destruction of a meteorite in the atmosphere.

Florenskiy's next expedition in 1961–62 concentrated on the study of the distribution of meteorite material in the soil. This was the first expedition to use a helicopter to deploy individual groups and transport heavy equipment. Thousands of tonnes of soil samples were collected at regular distances of 10, 20, 40, 60 and 80 kilometres from the epicentre. The analysis of these samples showed brilliant magnetic (magnetite) and glassy (silicate) globules, less than 0.1 millimetre in size. When a map of the distribution of these globules was drawn up, it showed that the globules occurred over a fairly well-defined ellipse, with high concentrations between 60 and 80 kilometres to the northwest of the epicentre. This pattern of distribution was probably the result of the wind direction of the day, which was from southeast to northwest. The appearance of the globules showed that they were formed in the atmosphere as molten matter condensed. Florenskiy was convinced that most of the globules were the remnants of a comet that exploded in mid-air.

In the early 1970s the Soviet scientist G.I. Petrov contended that as the Tunguska meteorite moved through the atmosphere, it rapidly evaporated and, when a large amount of vapour had amassed in front of the travelling body, it exploded and scattered in the atmosphere. This scattered matter then condensed into microscopic balls that settled on the vast Tunguska forest. Sphagnum bogs, which receive mineral nutrients only from the air, are most likely to assimilate such 'meteorite balls'.

As a result, Soviet scientists analysed the peat layers of the sphagnum bogs, which add a distinct layer of peat annually, and discovered fused silicate globules, up to 0.8 millimetres in diameter. The concentration of globules was substantially higher in the 1908 layer than in layers before and after the catastrophe. These globules were composed of rare-earth and heavy elements – the elements present in extra-terrestrial material. In his last paper published in English before his death in 2001, Academician Nikolai Vasilyev expressed doubts about the origin of the globules, as similar globules could also be formed during peat burning. 'The presence of small quantities of meteorite dust cannot be doubted', he wrote in *Planetary and Space Science*, 'but the problem of their connection to Tunguska remains open'.

Another attempt to study peat was made in the late 1970s by Emil Sobotovich and his colleagues at the Institute of Geochemistry and Mineral Physics in Kiev. After six years of investigation, they concluded that the Tunguska blast was caused by a 4,000-tonne stony meteorite that exploded before impact, scattering particles over a wide region. The researchers presented convincing evidence of their claim: a large number of tiny diamonds strewn over the Tunguska region. The team incinerated peat collected from the region of the fall in high-temperature ovens. In the ashes they found many irregularly shaped and extremely hard black grains. Further laboratory examination showed them to be diamonds.

Diamonds are formed under extraordinarily high pressures in volcanic pipes known as kimberlites, and since no such pipes have been found in the Tunguska area, the Kiev researchers concluded that the Tunguska diamonds were formed far from Earth (see the section 'The escapade of a gas' in Chapter Eight which contradicts this claim). Such diamonds are already known to exist in uralites, a class of meteorites, which presumably have been involved in deep-space impacts. These meteorites may originate ultimately from comets. Other researchers have suggested that the Tunguska object was a chunk of comet Encke, a periodic visitor to our skies. To support their claim that the tiny diamonds arrived in a meteorite, Kiev researchers cited high levels of radioactive carbon-14 in the peat. They said that such high levels of carbon-14 are found in meteorites that have been subjected to prolonged bombardment by cosmic rays in space.

Tunguska imprints in Antarctica

Globules collected by Florenskiy's 1961–62 expedition were enriched in iridium, a silvery metal that is abundant in extra-terrestrial bodies but rare on Earth, and contained other evidence of extra-terrestrial origin. Traces of the Tunguska fireball were not only present in the Tunguska soil; they have also been discovered in an Antarctic ice core. This startling revelation – made in 1983 by Ramachandran Ganapathy, an American expert on extra-terrestrial materials – provided fresh insights into the nature of the Tunguska object.

Ganapathy examined eight globules by a nuclear analytical technique with the aim of finding the answer to three questions: Are the globules truly extra-terrestrial? Are all the globules related to each other as would be expected if they originated in the Tunguska explosion? Can these globules be distinguished from the remnants of iron meteorites?

All eight globules contained iridium, nickel, cobalt, gold, chromium, antimony and iron. Their iron content ranged from 76 to 81 per cent, and they were all enriched in iridium, a reliable indicator of extra-terrestrial matter. One of the globules had a whopping 56,900 parts per billion iridium (rocks on the Earth's surface contain only about 0.3 parts per billion iridium; a meteorite may contain as much as 500 parts per billion). The concentration of nickel and cobalt, two elements always found with iridium in extra-terrestrial matter, was also present in cosmic proportions. 'There is no question about it', Ganapathy declared in *Nature*, 'the spheres are extra-terrestrials'.

The identical ratio of iridium and nickel in each globule also proved that they all came from the same extra-terrestrial body. The result also disproved the notion that the globules were from meteors falling continuously on Earth. The high abundance of chromium in some globules, however, indicated that the object was not an iron meteorite.

Ganapathy also reasoned that the eyewitness accounts of the Tunguska explosion, together with his findings, demonstrated that the Tunguska object was vaporised by the explosion in the atmosphere. The question of whether the debris from the event could have reached the stratosphere, and as a consequence been distributed globally, prompted him to search for debris in the Antarctic ice. Old ice and air bubbles trapped in it are precious scientific tools. They can illuminate the past, help answer questions about the present, and assist prediction about the future.

The average rate of accumulation of ice – about 7 centimetres per year at the South Pole – is a reasonably good chronometer for measuring time with depth. For his study, Ganapathy selected a 101-metre ice core drilled in 1974. A sample from a depth of 10.15 to 11.07 metres, which corresponded to 1908, contained four times more iridium than found in earlier years. Ice was similarly enriched in iridium during 1909 and the next few years. Then the iridium concentration dropped back to normal levels. This was the strongest proof that particles from the Tunguska fireball were scattered around the world by atmospheric currents. 'Because this iridium could only be deposited here by means of stratospheric fallout, it should be presented worldwide', he said.

The amount of iridium deposited in Antarctica could be used as a clue to calculate the total amount of atmospheric fallout from the event. The result: 7 million tonnes of debris. Ganapathy estimated that the object that exploded over Tunguska was a 7-million-tonne, 160-metre-diameter monster. He warned that his estimate did not make a distinction between comets and asteroids. However, he said that the object 'may well have been a stony asteroid'.

A year after Ganapathy's observations, Polish scientist Marek Zbik (now at the University of South Australia) examined 100 black magnetic globules from the Tunguska area. The globules varied both in size (from 7 to 350 micrometres) and shape (spherical, droplet-like, some even broken and damaged). Many of these globules turned out to be of terrestrial origin. The ratio of iridium and nickel in the remaining globules was very close to that observed by Ganapathy, proving them to be of extra-terrestrial origin. Zbik's analysis did not prove a definitive link between the globules and Tunguska.

The testimony of trees

Carbon atoms come in eight varieties, known as isotopes. Carbon-12 (6 protons and 6 neutrons in the nucleus) is the most common isotope. High-energy neutrons which continuously bombard Earth convert ordinary carbon-12 into radioactive carbon-14 (6 protons and 8 neutrons). Living things go on absorbing carbon-12 and carbon-14 until the time of their death. In the case of trees, carbon-14 is recorded in annual growth rings, which also give the age of trees.

It has been suggested that if the Tunguska explosion

had been caused by a comet, the hydrogen contained within it would have been compressed and heated as the comet passed through the atmosphere. Some of the hydrogen might have fused into helium, triggering a nuclear explosion that would generate high-energy neutrons and consequently carbon-14 in the atmosphere. Many scientists have measured carbon-14 in Tunguska tree rings, corresponding to pre- and post-Tunguska years. Examination of tree rings formed in 1908 shows a rise in carbon-14, but not enough to support the idea of an annihilation caused by a nuclear explosion. The rise in carbon-14 is attributed to the solar cycle, in which sun-spot numbers rise or fall over a typical period of eleven years. It has also been suggested that the burn-up of the Tunguska object in the atmosphere would have produced a temperature of a few million degrees, too low for nuclear reactions but high enough to produce carbon-14.

A groundbreaking experimental approach by a team of Italian scientists from the University of Bologna, headed by Giuseppe Longo, has uncovered new rem-nants of the Tunguska fireball. One of the team members, Menotti Galli, was on the 1989 Tunguska expedition, the first post-Cold War expedition open to international scientists. Galli, a physicist, is an expert on phenomena associated with cosmic radiation, including carbon-14.

During the expedition, Galli realised that the only witnesses to the 1908 blast still alive were the surviving trees. But their testimony was hidden in the resin formed around broken branches after the blast. Like amber, this resin could have acted as a trap for particles present in the atmosphere, including extra-terrestrial particles from the fireball. The resin would harden and form a protective coating around the branch. Eventually the resin would become enclosed within the growing branch. What Galli needed was to examine trees in the blast area. Their annual growth rings would point him to the 1908 sections, and if the fireball had showered any particles on the forest, they could still be intact in those sections of the trees. Bingo!

To collect the samples for examination, Galli and his colleagues, Longo, a nuclear physicist, and Romano Serra, an astronomer, attended the 1991 expedition. 'The Italians, accustomed to sipping espressos under Bologna's endless porticoes, found themselves slaking their thirst with brown swamp water laced with mosquito larvae', says Richard Stone in *Discover*, describing the 'ten difficult days' spent by the Italian scientists in Tunguska. With or without espressos, they still managed to collect resin deposited between 1885 and 1930 on fourteen branches of seven Siberian spruce trees abundant in resin. The trees were situated in different directions within a radius of 8 kilometres from the blast's epicentre. For comparison, they also collected resin from six branches of a tree growing at about 1,100 kilometres from the Tunguska site, and the roots of a tree blown down by the blast.

Back in Bologna, the researchers used a scanning electron microscope to examine their samples. In all they recovered 5,854 particles from the Tunguska branches and 1,183 particles from the two control trees. Their examination of these microscopic particles showed anomalously high abundance of iron, calcium, aluminium, silicon, gold, copper, titanium, nickel and other elements. Some of these elements are commonly associated with normal-density stony asteroids. This abundance peaked around 1908. Another interesting observation was that the smooth texture and spherical shape of particles from the Tunguska branches showed evidence of heating and melting. 'The blast wave would not have melted particles in the ground, where the conductivity was low', said Longo. 'That means the melted particles came directly from the cosmic body.'

Vasilyev pointed out in 1998 that the elements discovered by Longo's team in tree resin were similar to those found by Russian scientists in peat layers. This effect is most probably connected to the Tunguska body, he said, but for the final identification of the particles found in resin as the Tunguska matter, some additional corroboration is necessary, considering the fact that a large volcanic eruption in Russia on 28 March 1907 had produced a significant dust veil over the Northern Hemisphere for more than a year. 'There is no direct evidence that these materials have anything to do with the Tunguska body. On the contrary, there is good reason to believe we are dealing with fluctuations of the background fall of space dust.'

In the same year, Vladimir Alekseev of the Troitsk Institute for Innovation and Fusion Research in Moscow also expressed doubts on the methods used by Longo's team because of the presence of background particles which could come from volcanic eruptions. He came up with a different approach for hearing the testimony of trees: examining particles more energetic than those of the background. For his study, Alekseev selected a standing larch tree that had survived the Tunguska catastrophe. The tree, located near the epicentre, had a 10-centimetre vertical split in the stem. The split, according to Alekseev, could have been caused by shock waves which, coming from above, exerted a force on the growing tree. He took a wood sample from the split.

After removing resin from the sample, when Alekseev examined it with a high-powered microscope he noticed numerous solid particles up to 50 micrometres in size in the dense wood of the 1908 growth ring. The particles could be divided into four groups: metallic particles with jagged edges; spherical silicate particles; whitish particles; and black graphite-like particles. They had energies high enough to penetrate into the dense wood, and Alekseev was convinced that they were the remnants of the Tunguska body.

Based on the information obtained from the study of these particles, Alekseev proposed the following scenario for the Tunguska blast: the flight of the body was finished by multiple explosions and, therefore, the solid remnants are small particles. The multiple explosions could be responsible for the gunfire-like sounds repeatedly heard by the eyewitnesses of the event. There was a possibility of thermonuclear reaction on the surface of the body at the final stage of its journey in the atmosphere. Like some cosmic bodies, this body was probably enriched with deuterium (hydrogen-2). This deuterium could start a thermonuclear reaction which would convert deuterium to tritium (hydrogen-3). 'As tritium is involved in biological processes and it is radioactive, it can have genetic effects', Alekseev concluded.

During the second Italian expedition to Tunguska in 1999, the scientists not only continued their search for microparticles preserved in tree resins, they also looked for other remnants in the sediments at the bottom of Lake Ceko. This 500-metre wide and 47-metre-deep lake is 8 kilometres from the centre of the 1908 explosion. The group, which included Longo who had also attended the 1991 expedition, used an inflatable catamaran for the geological survey and for the coring operations. This work had two objectives: to check whether the lake is an impact crater of the 1908 event; and to detect mineralogical, chemical and biological evidence of the nature of the Tunguska fireball. Their experimental study showed that though the lake was formed by the impact of a cosmic body, it was definitely formed before the 1908 catastrophe. The core samples collected from the sediment have not yet conclusively been shown to be linked with the fireball.

Total ablation

Vladimir Svetsov of the Institute for Dynamics of Geospheres in Moscow removed a layer of mystery from the Tunguska event, when in 1996 he showed that the entire mass of the Tunguska body vaporised before it could reach the ground. Ablation – the loss of material from a cosmic body through evaporation or melting caused by friction within the atmosphere – of the Tunguska debris was total.

According to Svetsov, when a Tunguska-sized body enters deep into the atmosphere, it is broken into a large number of fragments, the maximum size of which is 10 centimetres. As the body decelerates, these fragments are separated from each other. Mathematical simulations show that stony fragments fully ablate either inside or outside the fireball due to high temperatures. All vaporised material does not reach the ground; it moves upwards in the atmosphere. Although Svetsov admitted that fragmentation processes were complex and certainly needed further investigation, he declared his scenario 'quite plausible'.

As for the Tunguska body, Svetsov said that it would have been heated to 15,000 degrees Celsius. This temperature was high enough to create an explosion 'quite comparable with that of a nuclear explosion'. Upon explosion the body broke into a vast number of fragments, typically 1 to 3 centimetres across but no larger than 10 centimetres. But the temperature was high enough to melt these fragments until nothing remained. Some of this microscopic debris condensed in the atmosphere and then was scattered over the Tunguska forest. 'The apparent absence of solid debris is therefore to be expected following the atmospheric fragmentation of a large stony asteroid', Svetsov concluded.

He also said that, similar to the 1994 impact of comet Shoemaker-Levy 9 on Jupiter, Tunguska debris was probably also widely scattered because of the turbulent wake of the asteroid. Svetsov also suggested that the microscopic particles recovered from the resin of trees by the Italian researchers 'might be recondensed material precipitated in the general vicinity of the impact site'.

Svetsov did leave some hope for Tunguska meteorite hunters: if some larger fragments accidentally gained significant speeds at altitudes between 15 and 20 kilometres, sizeable remnants could reach the ground. But they would have fallen to the ground at a distance of 5 to 10 kilometres southeast from the blast's epicentre.

Svetsov's calculations were based on the assumption that the Tunguska fireball was a 15-megaton asteroid-like object that hit Earth with a speed of 54,000 kilometres per hour and at an angle of 45 degrees. What if the object was a comet? Obviously, this analysis would then make no sense. NASA scientist Kevin Zahnle provides an argument that comet partisans would find hard to demolish. His argument goes something like this: all small impact craters on Earth are almost always produced by the relatively rare iron meteorites. The 1.2-kilometre-wide Meteor Crater in Arizona, for example, was produced by an iron body of essentially the same energy as the Tunguska explosion. The smallest known crater made by a stony meteorite is the 3.4-kilometre New Quebec crater. This raises a problem for the comet theory: if comets with energies of 15 megatons can reach the higher atmosphere before exploding, then the much more numerous asteroid, which most astronomers agree will penetrate deeper, should be cratering the land every 1,000 years. 'If Tunguska was a comet, where are all the Meteor Craters made by rocks?' Zahnle asks.

Environmental effects

During his first expedition to Tunguska in 1927, Kulik noticed quite rapid recovery of forest after the catastrophe. As we have seen, he wrote in his diary: 'The young twenty-year-old forest growth has moved forward furiously, seeking sunshine and life.' This accelerated growth of trees that survived the catastrophe has been studied by many Russian scientists. They have noticed that the effect does not coincide with the limits of the fire and destroyed forest, and is observed not only in surviving trees but also in younger trees germinated after the catastrophe.

It has been suggested that this accelerated growth is the result of genetic mutation caused by a nuclear explosion. Longo's team examined the abundance of carbon-14 in the 1903–16 tree rings, but found no traces of nuclear processes. This observation contradicted Alekseev's belief that radioactive processes were possible on the surface of the Tunguska body. But the Italian team suggested that the accelerated tree growth seemed to derive from such improved environmental conditions after the explosion as ash fertilisation by charred trees, decreased competition for light, and greater availability of minerals due to increased distance between trees.

A detailed analysis of the environmental effects of the Tunguska catastrophe was conducted by the American atmospheric scientist Richard Turco and his colleagues in the 1980s. Turco's analysis is based on the assumption that the Tunguska object was an icy comet nucleus, rich in water, ammonia, carbon dioxide and methane. As this object passed through the atmosphere, these substances contributed to the production of 30 million tonnes of nitrogen oxide. After comparing nitrogen oxide generated in nuclear bombs, Turco concluded that the Tunguska event might be compared approximately to 'a large-scale 6,000-megaton nuclear "war" in terms of nitrogen oxide deposited in the stratosphere'. In addition to this massive injection of nitric oxide in the stratosphere, the object also added about 1.5 million tonnes of water. This water helped in the formation of noctilucent clouds which caused bright nights. But the nitric oxide contributed to a longer and deadly effect. By a complex series of reactions, nitric oxide converts stratospheric ozone into oxygen. These reactions depleted the ozone layer that protects us from harmful ultraviolet rays.

The dust veil that hung over the stratosphere for years also contributed to climate changes. Turco said that about 1 million tonnes (Ganapathy's estimate: 7 million tonnes) of dust would be likely to decrease the average surface temperature by about 0.05 degrees Celsius. This resulted in an overall cooling of about 0.2 to 0.3 degrees Celsius in the Northern Hemisphere. Turco pointed out that the cooling trend might have been initiated by the 1907 volcanic eruption in Russia. Turco's team also studied weather records from the early 1900s and noted several other unusual weather conditions that appeared to begin around 1908 and lasted for several years: (1) an increase (above a decreasing trend) in the mean surface temperature over North America in both January and July beginning in 1909–10; (2) an increase in total Arctic ice between 1908 and 1911; and (3) a 50 per cent decrease from normal values in the number of tropical cyclones in the Atlantic and Caribbean Oceans.

Turco's team concluded its study by saying that 'this most impressive and consequential natural event' might have historical significance. Ozone depletion and climatic changes associated with large meteorites may have had a role in past events, such as the death of the dinosaurs 65 million years ago.

Sure, we now know about the environmental impacts of the Tunguska object and its remnants, but there remains the little matter of the identity of the object: was it a comet, an asteroid or something else?

CHAPTER SIX

THE INCREDIBLE JOURNEY
OF A BLACK HOLE

Mini black holes are still very much a figment of scientists' imagination, but they provide such a neat and fitting explanation for the Tunguska event that this theory has become the part of the folklore now associated with the mysterious fireball. In 1973 the American theoretical physicists A.A. Jackson IV and Michael P. Ryan Jr said that, since 'no crater and no meteorite material that can unambiguously be associated with the event have ever been found', a mini black hole could explain the Tunguska event.

In an article in *Nature*, Jackson and Ryan suggested that, after passing through the atmosphere, the mini black hole would have entered the Earth. Because of the rigidity of the rock there would have been no under-ground shock wave. With its high velocity, the black hole would have passed straight through the Earth in about fifteen minutes and exited through the North Atlantic, causing shock waves in the ocean and the atmosphere.

The mini black hole was much smaller than the full stop at the end of this sentence, but it had the mass of a large asteroid and quite a strong gravitational field for some distance from the body. As it passed through the air it became very hot, producing a deep blue trail of particles. Jackson and Ryan based their argument on the assumption that the damage caused by the Tunguska fireball was equivalent to a 2-megaton nuclear explosion. They estimated that the total energy in the black hole's blast wave would be within the same range.

Not a cosmic vacuum cleaner

A black hole is a star that has stopped twinkling. But why?

An ordinary star is one of the simplest entities in nature: it is a sphere of gas that is by mass 73 per cent hydrogen, 25 per cent helium and 2 per cent other elements. The temperature in the centre of a star is very high – high enough to fuse nuclei of hydrogen and helium together. The nuclear fusion produces energy that is radiated from the surface of the star as heat and light.

The universe has ten times as many stars as grains of sand on Earth – 70,000 billion billion stars (7 with 22 zeros after it), to be precise. To us all stars look similar, but no two stars are the same. Astronomers classify hundreds of billions of stars in our galaxy by their luminosity, colour, size and age. To us, stars also appear changeless. But stars are born, they live for millions of years, and they die.

The birth sites of stars are the dark clouds of gas and dust in our galaxy. The clouds, which are clumps of hydrogen atoms with a sprinkling of helium, are not uniform; they contain regions differing by density (1,000 to 10 million molecules per cubic centimetre) and temperature (–263 to –173 degrees Celsius), and regions with shapes ranging from spheroids to elongated tubes. Gravity tries to pull these clouds into the smallest possible space. Compression causes the gas to become hotter. Eventually the temperature and pressure rise high enough to ignite the gas. Hydrogen starts turning into helium, which creates vast amounts of energy. A star is born. All stars shine as a result of the nuclear fusion of hydrogen into helium, which takes place within their hot, dense cores, where temperatures may reach 20 million degrees Celsius.

Our Sun is a star – in astronomers' jargon, a main sequence star. A main sequence star – and 90 per cent of stars are these – fuses hydrogen nuclei into helium nuclei at its centre. The Sun has lived 4,600 million years as a stable star, and many billion years lie ahead. After consuming its hydrogen, the Sun will begin to expand. It will change into a type of star known as a giant, and will be about 100 times brighter than it is now.

After a few thousand years, the giant Sun will completely exhaust its supply of hydrogen and will shrink into a white dwarf – no larger than Earth, but so heavy that a teaspoonful of its matter would weigh thousands of kilograms. A white dwarf is so hot that it shines white-hot. Over billions of years, the white dwarf will turn black and cold. It will now be a dead star – a black dwarf.

A heavyweight star (a star with more than eight times the mass of the Sun) has a dramatic but brief life after becoming a supergiant. It expends its fuel so extravagantly that it collapses within a few million years. It then explodes as a supernova, which ejects an enormous amount of matter and even outshines the entire galaxy for a few days. The remaining matter forms a neutron star, only about 25 kilometres across, which contains tightly packed neutrons. These neutron stars do not glow, and are so heavy that even a pinhead of their matter would have a mass of a million tonnes.

Sometimes the crushing weight of a dying star like a neutron star squeezes it into a point with infinite density. At this point, known as singularity, mass has no volume and both space and time stop. The singularity is surrounded by an imaginary surface known as the event horizon, a kind of one-way spherical boundary. Nothing – not even light – can escape the event horizon. Matter falling into it is swallowed and disappears forever. That's why scientists call these regions of space-time black holes. If an astronaut passed through the event horizon of a black hole, gravitational forces would stretch his or her body into the shape of very long spaghetti, and when this very dead spaghetti slammed into the singularity of the black hole, the astronaut's remains would be ripped apart into atoms.

The radius of a black hole is the radius of the event horizon surrounding it. This is called the Schwarzschild radius, after the German astronomer Karl Schwarzschild who in 1916 predicted the existence of a dense object into which other objects could fall, but out of which no objects could ever come (the term 'black hole' was first used in 1969 by the American physicist John Wheeler; prior to that they were known as 'collapsars' or 'frozen stars'). The Schwarzschild radius is roughly equal to three times the weight of the black hole (in solar masses). A black hole weighing as much as the Sun would have a radius of 3 kilometres; one with the mass of Earth would have a radius of only 4.5 millimetres; and one with the mass of a small asteroid would be roughly the size of an atomic nucleus. A black hole's weird effects occur within 10 Schwarzschild radii of its centre. Beyond this rather limited distance, the only effect is through the black hole's normal gravitational pull. So, contrary to popular belief, a black hole is not like a cosmic vacuum cleaner that sucks in everything around it.

Not that long ago, black holes were in the realm of science fiction, but now there is convincing evidence for their existence. This evidence is still circumstantial – there is no way black holes can be observed directly. There are at least two species of massive black holes: smaller ones (a few times as massive as the Sun) that orbit normal stars; and their supermassive siblings (weighing many million Suns) which lurk in the centres of most galaxies. Our galaxy is believed to have a relatively small black hole that is as massive as 2.6 million Suns. A black hole with a mass 100 million times that of our Sun and a radius of 25 million kilometres squats at the centre of a galaxy 130 million light years away.

In 1971 the eminent theoretical physicist Stephen Hawking, who has greatly advanced our knowledge of black holes, proposed that during the first moments of the big bang that marked the birth of the universe, some areas were forced by the turbulence to contract rather than expand. This could have crushed matter into black holes that ranged in size from a few micrometres to a metre (their masses ranged from fractions of a gram to that of a large planet). This multitude of primordial or mini black holes may still exist, including some within the solar system, or even in orbit around Earth. These black holes have not yet been detected; there is not even circumstantial evidence for their existence.

Three years later, Hawking said that 'black holes are not really black after all: they glow like a hot body, and the smaller they are, the more they glow'. He proposed a mechanism by which black holes transform their mass into both radiation and particles that leave the hole. The result is that black holes gradually evaporate. So they do not last forever. The amount of radiation, now known as the Hawking radiation, escaping from a black hole is inversely proportional to the square of its mass; that is, the smaller the black hole, the shorter its life span. A primordial black hole with the initial mass of Mount Everest (and the size of an atomic nucleus) would have a lifetime roughly equal to the age of the universe, that is, 14 billion years; but a black hole with the initial mass of the Sun would vanish after about 100 million billion billion billion billion billion billion (1 with 62 zeros after it) years.

The Tunguska black hole

When Jackson and Ryan proposed their impeccably scientific black hole theory, they commented that many attempts had been made to explain the Tunguska event, 'ranging from the prosaic to the bizarre', and then suggested that 'a black hole of substellar mass such as those that have been postulated by Hawking could explain many of the mysteries associated with the event'. Their explanation has never been called 'prosaic' or 'bizarre', but it has certainly been described as 'imaginative and intriguing' by some scientists.

How can a small black hole explain the Tunguska event? Jackson and Ryan's case was based on three main arguments:

High velocity. The researchers assumed that the black hole had the mass of a large asteroid (about 100,000 billion to 10,000,000 billion tonnes), but its geometrical radius could be measured in micrometres. However, its gravitational field could be quite strong for some distance from the body. They also assumed that the black hole's escape velocity – the minimum speed an object must have to free itself from the gravitational pull of a planet or a star – was slightly greater than Earth's escape velocity, which is about 40,000 kilometres per hour. They calculated that, if the black hole began in interstellar space with zero velocity and fell freely to Earth's orbit, its velocity relative to Earth would be between 36,000 and 360,000 kilometres per hour. Thus, the black hole would travel through the last 30 kilometres of the atmosphere in about 1 second.

Bright blue 'tube'. The air around the passing black hole would heat to between 10,000 and 100,000 degrees Celsius. So most of the radiation from the shock front would be ultraviolet rays. The accompanying plasma column would therefore appear blue. 'These results agree well with eyewitness reports of the event and with measurements of the pattern of throwdown of trees at the site', Jackson and Ryan said.

No crater. 'Since the black hole would leave no crater or material residue, it explains the mystery of the Tunguska event', they said. 'It would enter the Earth, and the rigidity of the rock would allow no underground shock wave. Because of its high velocity and because it loses only a fraction of its energy in passing through the Earth, the black hole should very nearly follow a straight line through the Earth, entering at 30 degrees to the horizon and leaving through the North Atlantic.' At the exit point there would be another shock wave and disturbance of the sea surface. Jackson and Ryan suggested that oceanographic and shipping records could be studied to see if any surface or underwater disturbances were observed.

Scientists have found, forgive the pun, many holes in the black hole theory. 'The black hole would have shot straight through the Earth but unfortunately for the theory (although fortunately for us) the exit point, latitude 40 degrees, 50 minutes north, longitude 30 degrees, 40 minutes west, in the mid-Atlantic was not marked by an equally severe shock and blast wave', commented the British astronomer David Hughes in *Nature*.

Gerald Wick and John Isaacs of the Scripps Institution of Oceanography in California also wrote in *Nature*: 'Unfortunately, this miniature, hypothetical object cannot account for all the important phenomena known to accompany the event.' Their main argument centred around the small magnetic globules with high nickel content found in the Tunguska region. High nickel content confirms that the globules are of extra-terrestrial origin, but it does not necessarily ensure that they originated in the Tunguska blast.

'Most soil samples collected at random over the globe will contain similar cosmic dust', Wick and Isaacs said. 'The spatial pattern of the globules collected in the Tungus, however, shows that the cosmic dust likely originated from a massive meteorite body of dimensions vastly greater than a few angstroms.' However, they agreed that their discussion did not preclude the possibility that a black hole comprised the nucleus of the comet, or that black holes frequently might be the agents condensing the materials in such bodies.

William Beasley and Brian Tinsley of the University of Texas at Dallas claimed in *Nature* that several lines of evidence rendered the black hole theory extremely unlikely:

> *First*, many characteristics of the event indicated that the main part of the energy went into an explosion in the air. These characteristics include trees scattered on the ground, without branches or bark, in the direction opposite to the centre of the fall; an intense fire that seared trees; and in ravines, partially pro-tected trees that remained standing, but many with their tops broken. A typical meteorite buries itself below the surface and then dissipates its energy in an underground explosion. The Meteor Crater in

Arizona was produced in this way. 'No significant excavation was caused by the Tunguska explosion', remarked Beasley and Tinsley.

Second, a small cometary nucleus, consisting of a mass of frozen gases mixed with nickel-iron and silicate particles, would have a low degree of cohesion, and would fragment in the air and dissipate most of its kinetic energy before it reached the surface. A small black hole could produce a similar air blast, but would have passed through Earth in 10 to 15 minutes and caused a similar explosion at the point of exit, which would have occurred in the North Atlantic.

Third, about five hours after the Tunguska blast, six microbarographs in England recorded sound waves from the explosion. The approximate distance from the point of impact to the centre of the microbarograph stations is 5,720 kilometres, so the average speed of the waves was about 1,150 kilometres per hour, which is about the usual value for this type of wave. It is clear that the recorded waves were travelling from Siberia, and not from the North Atlantic. Sound waves from the site of the suggested exit explosion should, however, have arrived in England about three hours before the arrival of the Siberian wave. Beasley and Tinsley stressed that they had examined copies of the English microbarograph records, but had been unable to find any sign of waves from the North Atlantic exit point.

Fourth, a thick dust train along the path of the fireball immediately after its passage was noted by eyewitnesses. This observation is consistent with the deposition of the material in the atmosphere, rather than the loss of air into a black hole.

Fifth, exceptionally bright nights in Siberia and Europe imply that extra-terrestrial material was deposited in the upper atmosphere simultaneously with the impact. 'The wide area of atmospheric deposition is comparable to the dimensions of a cometary tail', Beasley and Tinsley said, 'and is not compatible with the idea of slow transport of dust vertically and horizontally from a ground level explosion'. This deposition of dust in the upper atmosphere could give rise to noctilucent clouds, which could account for the bright nights.

'All the evidence favours the idea that the impact which caused the Tunguska catastrophe involved a body with characteristics like a cometary nucleus, rather than a black hole', Beasley and Tinsley concluded.

A further challenge to the black hole theory came from the American scientists Jack Burns, George Greenstein and Kenneth Verosub. In the *Monthly Notices of the Royal Astronomical Society* they discussed inconsistencies in the predicted and observed thermal changes of soil and rock and seismic activity associated with the event. 'The point of entry of the hole into the Earth should be marked by a patch of melted and resolidified rock of diameter half to four kilometres, overlain by fused soil of comparable extent', they said. 'As the hole entered the soil it would have vaporised the water, oxidised the organic matter and fused the residual material such as quartz, feldspar and mica ... the point of impact should therefore be marked by a depression.' The Southern Swamp, or Kulik's Great Cauldron, is in fact a depression, but they pointed out that 'this depression may predate the Tunguska Event and is not inconsistent with other explanations'.

Burns, Greenstein and Verosub's calculations showed that Jackson and Ryan's black hole would release seismic energy equivalent to 1 million to 100 million megatons of TNT in the Earth, whereas the largest earthquake ever recorded (magnitude 8.3) released the equivalent of only 50 megatons. 'The absence of enormous seismic activity associated with the Tungus event therefore precludes its interpretation as a small black hole', they declared.

A postscript to Jackson and Ryan's popular theory appeared in Rupert Furneaux's book *The Tungus Event* (1977). Upon learning that no exit pulse had been found on the English microbarographs, the scientists were disappointed at the rejection of their theory: 'It begins to seem that the Tunguska event is more bizarre than any explanation put forward to date.'

Many decades have passed since the publication of Jackson and Ryan's black hole theory. We do now have a better understanding of black holes. Does this new knowledge support their theory? No scientific theory prohibits a wandering black hole striking Earth, but the question is: did a mini black hole pass through Earth on the morning of 30 June 1908?

CHAPTER SEVEN

THE MATTER IN QUESTION

The idea of the atom may have originated in Babylon or Egypt or even in India, but the story of matter started in 5th-century BC Greece with Leucippus and his pupil Democritus. They taught that matter was composed of empty space and an infinite number of tiny, indestructible particles called *atomos* or atoms. But Aristotle and other Greek philosophers preferred their 'elements' – earth, air, fire and water – out of which the whole world was created, and Democritus' idea was lost for two millennia. It was recovered and expanded in 1808 by John Dalton, a Quaker schoolmaster from Manchester, into his atomic theory.

The first real picture of the atom emerged in 1897 when the British physicist J.J. Thomson suggested that atoms are like a Christmas pudding, in which negatively charged electron 'raisins' are embedded in a spherical 'pudding' of positively charged protons. This delicious model was demolished in the early 20th century when Ernest Rutherford showed that the atom was like a miniature solar system with electrons orbiting around the central 'Sun' or nucleus composed of protons and neutral particles called neutrons. After announcing his model, the world-famous professor of physics at Manchester University, with a broad grin and in a booming voice, said to his close colleagues of his critics: 'Some of them would give a thousand pounds to disprove it.' No one had the temerity – or a thousand pounds – to challenge the model that soon, with some changes, became the icon by which we still recognise the atom.

While students struggled to understand the three-particle structure of the atom, physicists came up with complex quantum models of the atom and discovered an entire 'zoo' of elementary particles (so many, in fact, that it prompted Enrico Fermi to remark: 'If I could remember the names of all these particles, I would have become a botanist.'). The most famous of these particles are quarks, which were postulated in 1964 by the American physicist Murray Gell-Mann, who won the Nobel Prize in physics in 1969 for his work on them. Their name comes from a phrase – 'Three quarks for Muster Mark' – in James Joyce's novel *Finnegans Wake*. Until recently, quarks were considered the basic building blocks of matter, but some physicists now believe that quarks themselves are made up of even smaller particles. The physics of quarks and other elementary particles is very complex, but in simple terms we can say that each particle has three major characteristics: mass (some particles have zero mass); charge (every particle has a positive, negative or neutral charge); and spin (every particle spins like a top).

From matter to anti-matter

If you are a *Star Trek* fan you probably know that the starship *Enterprise* is powered by anti-matter. Anti-matter is not the stuff of science fiction; it does exist.

As early as 1898, Arthur Schuster, a British physicist, suggested the fascinating idea that an exotic type of matter could exist with properties that mirror those of ordinary matter. In a letter to *Nature* he wondered: 'If there is negative electricity, why not negative gold, as yellow as our own?' He added that this speculation was just 'a dream'. In 1928 the gifted British theoretical physicist Paul Dirac provided the mathematical basis for Schuster's dream. Dirac predicted that the electron, which is negatively charged, should have a positively charged counterpart: 'This would be a new kind of particle, unknown to experimental physics, having the same mass and opposite charge as the electron. We may call such a particle an anti-electron.'

The discovery in 1932 of the anti-electron (now known as the 'positron', short for 'positively charged electron') in the cosmic radiation by the American physicist Carl Anderson vindicated Dirac's bold prediction. Twenty-three years later, scientists at the University of California at Berkeley created the anti-proton in a particle accelerator. We now know that every fundamental particle has an anti-particle – a mirror twin with the same mass but opposite charge. The idea of anti-particles is now also applied to atoms – anti-atoms, which make up the anti-matter.

When anti-matter meets ordinary matter, they annihilate each other and disappear in a violent explosion in which mass is converted into energy as dictated by Einstein's famous equation $E = mc^2$, where E is energy, m is mass and c is the speed of light. The energy released in matter–anti-matter annihilation is awesome: in a collision of protons and anti-protons, the energy per particle is close to 200 times that available in a hydrogen bomb.

If matter and anti-matter annihilate each other, there is no likelihood of anti-matter existing on Earth, or even in the solar system. The solar wind, the spray of charged particles emitted by the Sun in all directions, would annihilate anti-matter. However, scientists speculate that anti-matter could exist in other parts of the universe, but so far they have found no evidence. This has not stopped them from creating anti-matter in the laboratory.

A team of scientists at CERN, the European particle physics lab in Geneva, did just that in early 1996. For about 15 hours they fired a jet of xenon atoms across an anti-proton beam. Collisions between anti-protons and xenon nuclei produced electrons and positrons. These positrons then combined with other anti-protons in the beam to make anti-hydrogen, the simplest anti-atom. Scientists could detect nine anti-hydrogen atoms. Hydrogen is the most simple (just one electron orbiting a single nuclear proton) and most abundant (it makes up about 75 per cent of the universe) of 114 chemical elements known to us. An anti-hydrogen atom would have a positron orbiting a single anti-proton. 'It's really the proof that there is an antiworld', exulted the CERN team's leader, Walter Oerlert of the Institute of Nuclear Physics in Germany. Since 1996, CERN scientists have been regularly synthesising anti-hydrogen atoms and have so far collected several hundred thousand of them. This harvest would provide scientists with an insight into the properties of anti-hydrogen.

So now there is the experimental proof that anti-matter does exist, what can it be used for? Because the annihilation of matter and anti-matter creates enormous amounts of energy – hundreds of times as much as generated in a nuclear reaction – it is tempting to look at anti-matter as a potential source of energy. This energy might one day provide the fuel for interstellar voyages, the same way matter–anti-matter annihilation powers the fictional spaceship *Enterprise*. The amount of anti-matter required for space flights is unbelievably small. A few hundred micrograms could fuel a spacecraft to Jupiter, and the round trip would take only a year.

If you find all this a bit too far-fetched, then what about the idea of an anti-universe – a universe parallel to ours. Enter it and you will find your anti-matter counterpart: anti-you. Don't shake hands – you'll annihilate each other.

Anti-matter Tunguska

In 1940, when the idea of anti-matter was nothing more than a mass of mathematical equations, the Russian scientist Vladimir Rojansky suggested the possibility of the existence in outer space of contraterrene meteorites (contraterrene, CT, or its phonetic transcription Seetee are obsolete terms for anti-matter, and were once very popular in science fiction; the ordinary matter was called terrene). Rojansky also said that such a meteorite 'would be entirely radiated away before reaching the sea-level'.

During the same year, *The New York Times* reported on 15 September: 'As the 22-foot cutter-type sailboat *Rockit II* was crossing Long Island Sound near Bridgeport, Conn., yesterday morning with four peaceful persons aboard, a shell screeched across her bow and exploded in the water 100 yards away.' A passenger recalled: 'It was a most disquieting experience ... The screech came first – an unholy noise. Then, a split second later, the explosion, about two points off the starboard bow. It blew up a great tower of water, twenty or thirty feet in the air. It was the strangest thing in the middle of the peaceful Sound. Why, there wasn't even a boat in sight! And not an airplane overhead!'

Authorities investigated the incident and found that no artillery shell could have exploded near the boat. As a meteorite on striking water would not explode, many astronomers of the time speculated that the explosion may have been due to the fall of a tiny contraterrene meteorite. In a comment on the *Rockit II* mystery in *Popular Astronomy*, Samuel Herrick Jr, an astronomer at the University of California, supported the contraterrene meteorite hypothesis and said that Dirac and other scientists 'are to be congratulated on one of the most ingenious (and entertaining) hypotheses of recent years'. He also warned his fellow astronomers that they 'will have to distinguish between the highly explosive fireballs or bolides from which no material reaches the ground, and which accordingly may be contraterrene, and those which are the source of terrestrial meteorites'.

This debate on contraterrene meteorites prompted Lincoln La Paz, a leading American meteorite expert who took a keen interest in Kulik's expeditions to Tunguska and co-translated many of his papers into English, to suggest in 1941 that the Tunguska meteorite was contraterrene in nature because of the great amount of energy released, the absence of impact craters, and the absence of nickel-iron positively attributed to meteorites. 'If a contraterrene iron meteorite of a size comparable to those of the largest irons conjectured to have fallen should strike the Earth', he said, 'an extremely powerful explosion would result, since, in addition to the large store of heat energy resulting from the transformation of the kinetic energy of motion of the meteoritic mass, a vast amount of energy would be liberated by its annihilation'. He pointed out that no original meteorite material would remain at the site of the explosion.

Herrick and La Paz's explanations generated a somewhat angry response from Harvey N. Ninniger, a noted meteorite expert and the President of the American Society for Research on Meteorites. He said that both phenomena could be explained by demonstrated facts without assuming the existence of any such 'purely hypothetical material', and 'we are surely courting a return to the days of "spirits and mystery" when we shrink from painstaking (or even back-breaking) investigations and seek refuge in untried hypotheses, especially when those hypotheses rest entirely on assumptions!'. (In 1928, Ninniger had urged American scholarly associations to send an expedition to Siberia 'to secure what is yet available of this greatest message from the depths of space that has ever reached this planet'. No one showed any interest.)

When scientists as distinguished as Willard Libby, who had developed the carbon-14 dating technique, and his colleagues Clyde Cowan and C.R. Alturi suggested in 1965 that the Tunguska object had been composed of anti-matter, they were probably not in danger of returning to the days of 'spirits and mystery'. Since Ninniger's warning in 1941 a lot had been discovered about anti-matter. Although its existence had yet to be experimentally proved, anti-matter was no longer considered a 'purely hypothetical material'.

In their detailed research paper in *Nature*, the three American scientists ruled out the possibility of a nuclear fission or fusion reaction and argued in favour of the anti-matter hypothesis. They said that neither fission nor fusion could explain the observed effects of the Tunguska explosion. To start a fission chain reaction (in which a heavy atomic nucleus splits into lighter nuclei), a critical mass of a fissionable material such as uranium or plutonium is required. The multi-megaton blast at Tunguska would require a large initial mass – well above the critical mass – which seems unlikely. On the other hand, fusion (in which lighter atomic nuclei combine to form a heavier nucleus) requires a sufficient amount of tightly packed deuterium that must be heated to several million degrees Celsius. Such a high temperature could not be obtained just by entry into the atmosphere.

The anti-matter hypothesis could explain the high nuclear energy yield of the Tunguska blast, but the researchers were quick to point out that 'several objections immediately arise' to this hypothesis. Two main objections were: (a) the lack of evidence for the existence of anti-matter; and (b) the anti-matter object would start disintegrating the moment it entered the atmosphere, and its largest yield of energy would be somewhere towards the middle of the path, rather than towards its end. 'A second look at the process tempers these conclusions, however', they said. Of the three models for nuclear explosion, they decided in favour of the annihilation of an anti-rock in the atmosphere. Their calculations showed that if the Tunguska explosion had been due to an anti-rock, it should have behaved like a 35-megaton fission or fusion bomb. The explosion would also have generated trillions of radioactive carbon-14 atoms.

As the feasibility of the American trio's anti-matter hypothesis depended upon the discovery of large amounts of radioactive carbon-14 in trees, they analysed carbon-14 content in sections of a 300-year-old fir tree that fell in 1951 in Tucson, Arizona, and an oak tree cut in 1964 near Los Angeles. They took nearly 90,000 counts of carbon-14 in tree rings from 1870 to 1930, which showed that the count peaked in 1909. However, the increase was much smaller than they had predicted. Their conclusion: although there are uncertainties, 'the data do yield a positive result'.

Recent measurements also show a rise in carbon-14, but not enough to support the idea of annihilation caused by a nuclear explosion, whether it be fission, fusion or anti-matter.

A decade after the publication of the anti-matter hypothesis, Hall Crannell of the Catholic University of America looked at other ways of measuring the anti-matter content of the Tunguska object. He said that silicon, and to some extent aluminium, are abundant elements in rocks, and when the Tunguska anti-rock hit the ground, ordinary aluminium was converted into radioactive aluminium-26. If the aluminium-26 content of rocks or soil is measured as a function of the distance from the centre of the explosion, he suggested, the highest concentration of aluminium-26 should be found near the centre. No one has yet carried out such measurements.

The British astronomer David Hughes rejected the anti-matter hypothesis on the ground that 'it is hard to understand how it penetrated to such a depth in the atmosphere and why the explosion maximised at the end of the trajectory and not midway along it'.

From anti-matter to mirror matter

The idea of a 'mirror world' was first suggested in 1956 by Chinese-American physicists Chen Ning Yang and Tsung Dao Lee. First, a tiny dose of particle physics, before you can enter their 'mirror world'.

The universe is held together by four types of fundamental forces – gravity, electromagnetism, the strong force, and the weak force – which are transmitted or 'mediated' by the exchange of elementary particles. The gravitational force, or gravity, is the long-range force responsible for the attraction existing between all matter: it holds you to the ground and Earth in its orbit. Its range is infinite. The electromagnetic force is the attraction and repulsion between charged particles: it enables a light bulb to glow and a magnet to stick to your fridge. Its range is also infinite. The strong force is the 'glue' that holds together an atomic nucleus: it binds quarks to make protons and neutrons. The weak force is also a kind of nuclear force: it causes elementary particles to shoot out of the atomic nucleus during the radioactive decay of elements such as uranium. The range of the strong and the weak force is extremely short. The electromagnetic, the weak and the strong forces are very similar and are very well understood by physicists, but gravity is still a mystery, and little is known of its relation to the other forces.

The existence of anti-matter leads to the idea of symmetry, that is, every particle has a mirror-like twin. An anti-particle would look just like the ordinary particle, except that left would be switched with right. Physicists call it reversing the parity (parity is just a sexy word for left–right or mirror symmetry). Symmetry also applies to laws of physics such as the rules governing the interaction of elementary particles. All the original laws should continue to work in the same way: whatever could happen in the real world would also happen in the anti-matter world.

But nature's symmetry is flawed. Certain interactions of elementary particles always produce a particle always spinning in the same direction. For example, when an atom emits a neutrino it always spins in the same direction – left-handedly (if it were coming towards you, you would see it spinning clockwise). Reflected in a mirror, however, a neutrino would be right-handed (it would always spin anti-clockwise). In contrast, electrons can spin in both directions. As many elementary particles display a preference for left over right, the universe seems left-handed. Why? Physicists do not know.

In 1956 Yang and Lee suggested that the evidence for left–right symmetry was weak in interactions involving the weak force (which led Wolfgang Pauli, who had dreamed up neutrinos in 1930, to lament: 'I cannot believe God is a weak left-hander.'). This prediction was soon confirmed experimentally by other physicists. Mirror symmetry or parity was now dead. Asymmetry was the new king. The discovery of asymmetry won Yang and Lee the Nobel Prize in physics just a year later.

Like nifty accountants, Yang and Lee had to balance the books. They proposed a way to restore perfect left–right symmetry to nature: every right-handed particle might have a left-handed particle, and vice-versa. This means that in addition to the anti-matter world, there might also exist a mirror world. In the mirror world, all neutrinos would be right-handed. Considered together, the real world and the mirror world would restore the symmetry that appears to be lacking in each.

Welcome to the mirror world – a world of mirror planets, mirror stars and even mirror life, all governed by mirror forces. This world is as fanciful as the one Alice entered *Through the Looking Glass*.

In this world, particles are right-hand or mirror images of ordinary particles. They also have the same mass as their ordinary counterparts. Thus, one force that acts on both ordinary matter and mirror matter is gravity. But there would not be any interaction between ordinary matter and mirror matter through nature's other three forces – the electromagnetic, the strong and the weak. We should be able to detect gravitational force when mirror matter comes near ordinary matter. The detection of this force would betray the presence of invisible mirror matter. The testability of this idea takes mirror matter out of the realm of science fiction into reality.

Because we are made of ordinary matter, we can neither see nor smell mirror matter (or our mirror matter twins, even if they were dressed in their brightest mirror matter clothes and soaked in mirror matter perfume). If you did encounter your mirror matter twin, you would pass right through him or her. You would also be invisible to your twin.

No mirror matter has yet been discovered or made in the laboratory, but neutrinos provide a misty glimpse of the mirror world. Neutrinos are the most pervasive elementary particles in the universe. There are about 50 billion neutrinos for every electron; they are everywhere but they cannot be seen and rarely interact with matter. Tens of thousands pass through our body every second. They have no charge and, although previously thought to have no mass at all, they are now believed to have a small amount of mass. There are three known types of neutrino – muon, tau and electron – and they are all created in the centre of the Sun, in supernovas and in the cosmic rays hitting the upper atmosphere. (In his famous book *The Quark and the Jaguar*, Murray Gell-Mann writes that the neutrinos produced by the Sun 'reach the surface of the earth by raining down on us during the day, but at night they come up at us through the earth'. This aspect of neutrino behaviour inspired writer John Updike to write a poem entitled 'Cosmic Gall'. An excerpt: 'The earth is just a silly ball / To them, through which they simply pass, / Like dustmaids down a drafty hall / Or photons through a sheet of glass.')

Physicists have calculated the number of electron neutrinos that should reach Earth from the Sun. But they have actually detected fewer of these than predicted. Recent experiments have shown that neutrinos can change from one type to another. Some types of neutrinos are not spotted by neutrino detectors, which explains the discrepancy. The proponents of mirror matter solve the puzzle of missing solar neutrinos by suggesting the existence of a fourth type of neutrino – mirror neutrinos. These are so ghostly that they don't make their presence known to bewildered physicists.

There is more good news for those who believe in the existence of mirror matter. And it comes from a particle called orthopositronium. Positronium is like a hydrogen atom, but instead of an electron orbiting a proton, an electron orbits a positron, its anti-matter counterpart. If the spin of the electron and the spin of the positron point in the same direction, the atom is known as orthopositronium. In 1986, Harvard physicist and Nobel Laureate Sheldon Glashow suggested that orthopositronium could oscillate between mirror and ordinary orthopositronium – jumping back and forth through the mirror.

Orthopositronium is ephemeral; it lasts a mere 142 nanoseconds before its components annihilate each other in a burst of tiny energy in the form of three undetectable photons. However, in the 1990s, when physicists made a batch of orthopositronium, they found that its lifetime is shorter than 142 nanoseconds. In 2000,

Robert Foot of the University of Melbourne and Sergei Gninenko of CERN suggested that mirror ortho-positronium could explain the discrepancy. This could be due to orthopositronium changing fleetingly into its mirror matter form and then back again. The mirror orthopositronium would go undetected, and that could account for the shorter lifetime measurements.

Although mirror matter is expected to interact with ordinary matter only through gravity, recent experiments suggest a small electromagnetic attraction between mirror and ordinary particles. This coupling probably comes from the tiny electric charge that mirror electrons and protons are believed to have. This charge is about a millionth that of their ordinary counterparts. The tiny electromagnetic interaction between mirror particles and ordinary particles, if it exists, has interesting implications. It would make mirror stars visible if they had some embedded ordinary matter. This interaction would also be sufficient to heat up a body of mirror matter if it entered Earth's atmosphere. And that's where Tunguska enters the mirror world.

Mirror matter Tunguska

Robert Foot, who has been studying mirror matter since 1991, became interested in the Tunguska event when in 1999 he watched the television documentary *As It Happened: The Day the Earth Was Hit*. He became convinced that the event was not fully understood by scientists and that they were ignoring the crucial evidence such as the funnel-shaped holes discovered by

Kulik. He also thought it most unlikely that an ordinary matter asteroid or a comet could completely disappear in the air without leaving any traces, however minute.

In 2002, Foot proposed an interesting solution to the Tunguska puzzle. In his book *Shadowlands: Quest for Mirror Matter in the Universe*, he suggested that the event was caused by a mirror asteroid. As it dived into the atmosphere the heat caused it to explode at high altitude. The explosion caused a shock wave that wreaked havoc on the Tunguska taiga but didn't leave a trace of an impact crater. He estimated that the mirror matter space-body was roughly 100 metres in size and weighed about 1 million tonnes. Such a heavy (ordinary or mirror) body would not lose much of its velocity in the atmosphere if it remained intact. However, if it were to break up for any reason, the energy of the body would be rapidly dumped into the atmosphere, leading to a huge explosion.

As for the nature of the space-body, Foot said that it was most likely made from mirror matter ices, such as mirror H_2O ice. An important difference between the mirror ices and the ordinary ices would be that the mirror ices would not be melted by light from the Sun, and therefore would be likely to be relatively abundant in the inner solar system. On entering the atmosphere, a mirror H_2O ice body would vaporise during the flight, and any leftover fragments would eventually melt after striking the ground. 'This could explain why no substantial mirror matter fragments were found at Tunguska; most of the space-body had vaporized after it exploded in the atmosphere, any remaining fragments had melted by the time Kulik arrived there', he said. 'Once in the liquid state, mirror matter should seep into the ground, probably making its extraction impossible.'

However, Foot has left some hope for Tunguska trophy hunters. The mirror body might have some embedded amount of ordinary matter, so a tiny amount of ordinary extra-terrestrial material was possible. Also, any mirror matter fragments that survived and hit the ground could potentially cause small craters or holes. 'Perhaps the most interesting facet of this interpretation of the Tunguska event is that there should be large pieces of mirror matter still lodged in the ground at the Tunguska site', he said. 'Nobody has looked.'

Perhaps the most spectacular way to test Foot's idea is to actually find mirror matter in the ground at Tunguska. 'Any mirror matter fragments would have melted when they hit the ground and reformed becoming mixed with ordinary matter at some distance underground', he said. 'There may be some amount close to the surface which could potentially be extracted and purified.' He suggested that the mirror matter might be separated from the ordinary matter in a laboratory centrifuge. But there is a catch: once the heavier mirror matter had separated from the ordinary matter, it would fly out of the centrifuge test tubes. As you can't see it, you can't catch it. The experiments, however, would prove the existence of mirror matter if the mass of the test tubes and their contents after the experiment were less than they were before. The missing mass would be the mass of the mirror matter you failed to catch.

'It would be a very exciting experiment and lots of fun

too!', Foot believes. If you were ever interested in making a name for yourself as an experimental scientist, here's your opportunity. However, you must heed Foot's warning before you pick up your shovel and head to Tunguska: 'It is possible that mirror matter could be hazardous to health.' He takes no responsibility for any cases of mirror matter poisoning.

Alice probably knew about mirror matter poisoning. Just before she stepped through the looking glass, she asked her cat: 'How would you like to live in a Looking-glass House, Kitty? I wonder if they'd give you milk in there? Perhaps Looking-glass milk isn't good to drink.' Now we know that this milk would be made of mirror molecules and perhaps it wouldn't be good to drink, for Kitty at least. Mirror-Kitty would love it, for sure.

There's another matter

And it's called quark matter. The incredibly tiny quarks come in six types or 'flavours': up, down, strange, charmed, bottom and top. Protons and neutrons are made from up and down quarks. Other quarks are not found in ordinary matter. Scientists, however, believe that a strange quark matter – a form of matter made of up, down and strange quarks – was formed in the big bang that marked the beginning of the universe some 13 billion years ago. This matter is so dense that a teaspoonful of it would weigh billions of tonnes.

In 2002, the orbiting Chandra X-Ray Observatory spotted a star which scientists believe is a quark star. Theorists have long suspected the existence of quark stars, collapsed stars that are denser than neutron stars but not dense enough to become black holes. The observed star has a radius of 5 to 6 kilometres. This radius, scientists believe, is about half of what would be expected if the object were a neutron star, but about right if it were a quark star. If the strange quark matter does really exist, it could destroy ordinary matter by converting protons and neutrons to quarks. This process could spread like cosmic wildfire through space.

In the same year, a team of researchers from Southern Methodist University in Dallas said that not only does the strange quark matter exist, but it passed through Earth twice in 1993. The first event occurred on 22 November, when an object entered Earth off Antarctica and left the Indian Ocean south of Sri Lanka 26 seconds later. In the second event, on 24 November, an object entered south of Australia and exited near Antarctica 19 seconds later. These entry and exit points imply that both objects were travelling with a speed of 144,000 kilometres per hour. Both events were recorded by several monitoring stations but no satisfactory explanation for the events has ever been suggested.

In 1984 Sheldon Glashow said that strange quark matter would pass through Earth with a dramatic effect: a 1-tonne object would release 50 kilotons of energy, which would be spread through Earth along the path of the object. The Southern Methodist researchers began looking for such events in 1993 and selected the two events mentioned above from more than a million records collected by the US Geological Survey between 1990 and 1993 which were not associated with traditional seismic disturbances. According to Eugene Herrin, one of the team members, normal earthquakes and explosions are point sources – the energy radiates from a single point. When a clump of strange quark matter – known as a strangelet or strange-quark nugget – passes through Earth, you have a linear source with energy radiating from the whole of the line through Earth, and this would give a different pattern in the way seismic stations record the data. The 1993 events – caused by strangelets just one-tenth the breadth of a hair and weighing nearly a tonne – left a distinct linear quark matter pattern. 'We can't prove that this was strange quark matter, but that is the only explanation that has been offered so far', Herrin said.

The impact of strangelets on an inhabited area would probably be less violent than that of a meteorite. 'It's very hard to determine what the effect would be', said Herrin. 'There would probably be a tiny crater but it would be virtually impossible to find anything.'

Did a quark matter rock strike Tunguska on 30 June 1908? There is a possibility, but no one has the definitive answer, yet.

CHAPTER EIGHT

A BLAST FROM BELOW

The line-up of suspects in the crime against 60 million trees of the Tunguska taiga so far includes a comet, an asteroid, a mini black hole and a rock of anti-matter or mirror matter. Some contemporary scientists reject these extra-terrestrial suspects on the grounds that no definitive ET remnants have ever been found. They point the finger at something much closer to home. Their line-up of terrestrial suspects includes a massive gas explosion, a giant lightning ball and a so-called geometeor.

The escapade of a gas

Wolfgang Kundt, a professor of astrophysics at the University of Bonn, dismisses comet and asteroid theories as pseudo-science and suggests an alternative 'volcanic blow-out' scenario for the fireball. As natural gas escaped from narrow underground volcanic vents, it became charged. The escaping gas, which contained mostly methane, raced upwards at high speed and started mixing with air. After a few hours, this charged, volatile mixture sparked lightning that ignited it into a fireball. This massive fireball contained as much as 10 million tonnes of natural gas, and caused the devastation we all know about.

154

The Tunguska site, according to Kundt, lies in the middle of the 250-million-year-old Kulikovskii volcanic crater, which has several faults or fractures running through layers of rocks. In the Tunguska blast, the gas escaped through a kimberlite, a carrot-shaped volcanic pipe in the rock, which had been formed when hot magma from the Earth's molten mantle pushed upwards under high pressure. Diamonds are formed in kimberlites more than 150 kilometres beneath the surface, and brought up in volcanic vents or pipes called diatremes. Kimberlites are named after Kimberley in South Africa, where legendary diamond reserves were found in the 1870s. Once thought to be common only in South Africa, kimberlites have now been found in other parts of the world.

The first expedition to the Tunguska site was carried out in 1910 by a wealthy Russian merchant and goldsmith named Suzdalev who, on return, urged the locals to keep silent about his expedition. Apparently, they obeyed. We do not know whether Suzdalev left Tunguska with his sledge filled with sacks of diamonds, but we do know from Kundt that the ejection of gas caused the formation of diamond-rich kimberlite diatreme pipes.

Kundt claims in the journal *Current Science* that dozens of funnel-shaped holes, including the famous Suslov crater discovered by Kulik, had been 'blown from below' during the morning of the explosion. He cites Evenki eyewitness accounts in support of his claim. He explains the presence of ice crystals in permanently frozen mud in the Suslov crater by saying that 'during its formation water should have intruded into its cavities'.

Russian scientists have explained this phenomenon as the result of permafrost, and have pointed out that similar holes are often found in other parts of Siberia.

Kulik had also found a decayed tree stump at the bottom of the Suslov crater. How could a tree stump make its way to the bottom of a crater blown from below? Kundt explains: 'There have been dozens of trees, standing on top of what is now the Suslov hole. Most of them were hurled to large distances, but one of them may have managed to fall back in, more or less upright. Alternatively, this stump could have slid back from the crater's rim along with transient mud flow.' Kundt believes that his theory has also answered the questions: 'Why did several expeditions find large numbers of detached tree stumps laying around in the Great Cauldron and its vicinity?'; and 'How did they get there?'. He says: 'To me, they are a clear indication of ejections, from the holes at whose surfaces they had grown.'

The pattern of treefall in the blast area, which has been studied by many researchers, has many unique features. The almost radial pattern – 'wiggly rather than straight-line radial', according to Kundt – has five centres and follows the valleys and hills. It shows islands of survived trees in the valleys. It also has 'telegraph poles' near the epicentre, which are reminiscent of the Hiroshima blast waves. Kundt claims that such fine structures of blasting and felling could not be explained by one big explosion on the ground. He considers the impact models inconsistent, as 'all of [them] produce parallel treefall patterns, if properly evaluated'. The actual pattern 'requires several successive localized explosions near the ground', he says.

A detailed study of the treefall pattern made by the 1991 Italian expedition has also suggested more than one centre of the explosion.

Kundt explains 'bright nights' observed in parts of Europe and Asia after the Tunguska explosion by saying that 'such nights are unique in the recorded history except for the 1883 Krakatoa volcanic eruption'. He says that the bright nights of Krakatoa and Tunguska were due to natural gas – mainly methane – which could rise to an altitude of 200 kilometres, where it was reheated by solar radiation and by slow burning with the surrounding atomic oxygen and rose again to an altitude of about 600 kilometres. Water vapour formed during slow burning froze out as snowflakes, which scattered the sunlight and gave rise to bright nights.

Kundt's other reasons against the extra-terrestrial origin of the Tunguska blast include:

- An asteroid would have left a trace, whereas a comet would have exploded too high, and also would have been discovered before the impact.
- The absence of any remnants of the interstellar body. An iron comet would have left a large, lasting crater. A stony asteroid would have left either big fragments or at least 4-millimetre-thick dust.
- Several witnesses reported the sound of gunfire before they saw a 'pillar of fire' in the sky. This order of events is expected from a volcanic blow-out, not from an extra-terrestrial impact.
- The heat felt by many witnesses at Vanavara, about 70 kilometres from the explosion site, cannot be

explained by a meteorite trail. A meteorite trail cannot produce such intense heat because it is narrow; it would have to pass quite near to Vanavara with a rather large speed. 'What counts is the spherical angle of the hot source, seen by your face: you feel the heat of a near bonfire – covering a large spherical angle – but you cannot possibly feel the heat of a (short and narrow!) meteoritic trail', he says.

For his last reason – number 19 on his list – Kundt relies on statistical odds. It is well known to geologists that only a small number of terrestrial craters were gouged in by rocks from space; most of them were formed by volcanoes. In all the faces of volcanism – ranging from hardly noticeable outgassing, through lava flows, to mud volcanoes, real volcanoes and explosive, supersonic ejections – rising natural gas is the primary piston.

There are other reported cases of natural gas explosions, but none as dramatic as the Tunguska blast. In his *Current Science* paper, Kundt quotes a 1988 incident reported to him by the American geologist Thomas Gold: 'A United Airlines plane on the way from Tokyo to Honolulu in calm air experienced a sharp upward bump followed, in a fraction of a second, by a mightier downward movement with a recorded speed implying a downward acceleration of 4 g.' Gold explained the upward bump as the crossing of a methane cloud rising at high speed. The plane's engines then ignited the methane–air mixture above the plane. This explosion forced the plane downwards and injured many people seriously. The plane had to return to Tokyo to attend to the wounded.

Kundt considers Tunguska 'a lovely detective story which requires a broadly educated physicist's mind for its resolution'. He says that he owes a lot of thanks to Andrei Ol'khovatov (see 'Ghostly geometeors' below), whose sober analysis converted him from the mainstream opinion to a physically consistent one.

After Kundt's presentation at a conference on environmental catastrophes in London in 2002, Jesus Martinez-Frias of the National Aerospace Institute in Madrid (obviously with a 'broadly educated physicist's mind') said that Kundt's geophysical hypothesis was 'a fresh idea'. 'It could be the answer', he suggested.

Like Kundt, Vladimir Epifanov, a geologist from the Siberian Research Institute of Geology, Geophysics and Minerals, also believes that the epicentre of the Tunguska explosion is indeed located just above a major oil and gas field. But he suggests a different mechanism for the explosion: a powerful fluid jet that had suddenly shot up under high pressure from the depths of the Earth.

According to Epifanov, gases from the oil deposits and methane from the coal beds, which had accumulated under a thick cover of basalt, suddenly broke through one day. A moderate earthquake could have initiated the process. The fluid jet was accompanied by dust that created a layer of aerosols in the upper atmosphere. If this layer became charged with electricity, it could have produced the spark that set off the explosion. The fireball then rushed towards the ground, flattening trees in a circular pattern for many kilometres. However, the cooling as the escaping gas rapidly lost pressure could have formed an ice dome around the place where the gas discharged, protecting the trees in the centre of the blast. This would not be the case with an impact from space.

Critics of the 'blast from below' scenario ask: what about the fireball that was seen by many witnesses racing across the Siberian sky from the south-southeast to the north-northwest just before the explosion? A volcanic gas blast fails to explain a racing fireball in the sky. Have these eyewitness accounts flown out of the window – just like a ball of lightning?

A giant lightning ball

A few years ago, a reader wrote to the 'Science Times' section of *The New York Times* about a lightning ball that was seen by her family to 'enter the glass front door, go right past us (or possibly even through us) in the living room and leave by the back window, where it hit a tree, causing some damage'. The Science Times commentator joked, 'Next time, take a picture', because it is one of nature's rarer phenomena and few photographs of it exist. It is also the least understood. Ball lightning has attracted the attention of scientists for two centuries, but it remains an enigma – dismissed by many as a myth or an optical illusion.

Over the years, scientists have collected thousands of accounts of sightings of ball lightning. In 2002 the Royal Society's journal *Philosophical Transactions* presented a selection of recently reported sightings. One account describes a lightning ball as it entered through an open window in the pantry of a house in Johannesburg: 'It entered the kitchen around the corner then sped out of the kitchen again round another corner and into the passage and the hall where it hit the tin bucket with a clang! Certainly when we ran to check, the bucket was too hot to pick up and its paint had blistered!' In another account, a white-grey lightning ball 70 to 80 centimetres in diameter and with the glow of an incandescent lamp of 200 watts bounced on the head of a Russian teacher who was with her friends: 'It appeared as if from nowhere. We got frightened, squatted, and connected our heads, creating a circle. The ball suddenly began to move over us in a circle, and it also moved up and down. It was at a height of 0.5 metre above the ground. Then it "chose" my head and began to jump on it, up and down, like a ball. It made more than 20 jumps. It was as soft as a bubble.' The journal also listed an extraordinarily large – about 100 metres in diameter – lightning ball that was caught on colour film by a park ranger in Queensland, Australia. It was anchored to the ground and lasted surprisingly long, about five minutes.

One of the rare accounts of ball lightning witnessed by a respected scientist comes from the British radio astronomer R.C. Jennison, who encountered a lightning ball on a late-night flight in stormy weather in 1963. He described it in a letter to the journal *Nature* in 1969 as a blue-white glowing sphere a little more than 20 centi-metres in diameter which emerged from the pilot's cabin and passed down the aisle of the aircraft approximately 50 centimetres from him, maintaining the same height and course for the whole distance over which it could be observed. It then passed through the metal wall of the aircraft.

From such accounts, scientists have painted a picture of this bizarre phenomenon, which is always observed during stormy weather. A lightning ball is usually seen as a free-floating, luminous sphere that shines for a few seconds to a few minutes before it either explodes with a sharp bang or flicks out in silence. It can be almost any colour, sometimes even a combination, but green and violet are rare. Its size varies from a small ball to a giant globe several metres in diameter. It may suddenly appear in the air, or even from holes in the ground, chimneys, sewers and ditches. It usually moves horizontally in the air (at speeds between 3.5 and 350 kilometres per hour) about a metre above the ground, but can climb utility poles and then dart along power or telephone lines. It can even dive down chimneys and squeeze through spaces much smaller than its size, but it never changes its size. It seems cool to the touch, but it may destroy electrical equipment, melt glass, ignite fires and scorch woods or singe people and animals. Sometimes a hissing or crackling noise can be heard. It may leave behind a sharp and repugnant smell, resembling ozone.

Ball lightning has been the subject of serious scientific research since the early 19th century, but no consensus theory has yet emerged. One of the popular theories is the plasma theory which says that a lightning ball is a sphere of plasma, or a hot gas of electrons and positively charged ions. Another theory that is gaining favour comes from New Zealand scientists John Abrahamson and James Dinniss. When lightning strikes soil, it turns silica in the soil to pure silicon vapour. As the hot vapour cools, the silicon condenses into a floating ball of silicon aerosol held together by electrical charges. The chemical energy stored in silicon is slowly released as heat and light. Because the ball would become visible only over the latter part of its lifetime, it would appear to materialise out of thin air after a lightning strike. So simple, yet so amazing! Unfortunately, most of us will never see it (about 1 per cent of the population will see ball lightning in a lifetime).

Like Tunguska, ball lightning is a favourite of both scientists and charlatans. Is there a link between Tunguska and ball lightning? Key in 'ball lightning + Tunguska' in Google's search engine and you will be instantly presented with a list of thousands of web pages. A close analysis shows that most of these are alternative science pages discussing Tunguska and ball lightning in a similar vein as UFOs and alien abductions. The scientific link between Tunguska and ball lightning seems tenuous.

In his book *Cauldron of Hell: Tunguska* (1977) the American science writer Jack Stoneley poses the question: could some particularly massive form of ball lightning conceivably be associated with the Tunguska event? To answer it, he quotes the British scientist Anthony Lawton, who was also the scientific editor of Stoneley's book, as saying that to cause such devastation would require a lightning ball nearly 1 kilometre in diameter. Stoneley claims that from eyewitness reports early Tunguska researchers reckoned the fireball to be about 1 kilometre across. 'This is so close that we cannot dismiss the possibility that the Siberian monster might have been a giant lightning ball', he writes.

Can a lightning ball grow to such an enormous size? 'Lawton maintains it is possible, provided the ball lightning is composed in a particular way', Stoneley writes. 'He suggests that if the fireball is made up from dust particles bound tightly together by an electric charge, it could achieve these huge dimensions.' Lawton's prediction now sounds believable in view of the new aerosol theory of ball lightning.

David Turner, a British physical chemist now retired and living in the United States, looked at Tunguska and ball lightning in a different light. In an exhaustive analysis in *Physics Reports* he says that various studies of Halley's comet in 1986 found the temperatures in the plasma region of the comet to be much higher than expected. This looks like an exact analogy of the high temperatures implied for lightning balls, he says. He then went on to say that these observations could have relevance to 'one of the most spectacular and perplexing events' in the 20th century.

Turner lists five factors that do not support the asteroid theory:

- The very high percentage (more than 10) of the energy released as electromagnetic radiation.
- The occurrence, six minutes after the explosion, of a local magnetic storm that lasted more than four hours.
- Optical anomalies (bright nights and so on) seen in some parts of Europe and Asia, which began a week before the event but peaked on the morning of the explosion.

- The failure to find virtually any fine material on the ground that can be reliably associated with the impacting body.
- An apparent change in direction of the falling body (both in the horizontal and the vertical plane) which occurred shortly before the explosion.

For Turner the most important outstanding question is the determination of the maximum temperature sustainable within ball lightning. Present estimates vary from 400 to 15,000 degrees Celsius. He believes that this is closely related to the question of whether the Tunguska event could have resulted from a natural hydrogen bomb explosion. He is implying that the plasma in the comet, which was like a lightning ball, was hot enough to start a hydrogen bomb reaction. 'It may be premature to discount the comet hypothesis at this stage', he advises.

G.G. Kochemasov of the Russian Academy of Sciences believes that partisans of the comet and asteroid theories fail to consider two important points. First, the anomalous atmospheric conditions long before the event. Second, the non-linear motion of the object. These two reasons are similar to those listed by Turner. Kochemasov says that a giant lightning ball can explain these two anomalies. The Earth's restless ionosphere causes various electrically charged events such as aurora borealis. As flights of ball lightning have been noticed along geomorphological boundaries, it is possible that a giant lightning ball formed in the Tunguska region, which is in an area of volcanic and tectonic activities. Thus, it was a home-grown product without any ET connections.

Kochemasov estimates the diameter of the lightning ball to have been about 200 metres. Such giant balls have never been observed but, he believes, we have to think in terms of thousands of years, or geological timescale. He cites two historical occurrences of giant ball lightning: the archaeological evidence suggesting that the ancient Indus Valley city of Harappa (now in Pakistan) was ruined after an enormous fire; and mention in the ancient Indian epic *Mahabharata* of an 'explosion' that caused 'dazzling light, a fire without smoke'.

Kiril Chukanov, an independent researcher on ball lightning in Salt Lake City, Utah, also believes that the Tunguska fireball was an enormous lightning ball about 500 metres across. On his 'Chukanov Quantum Energy' website and in his self-published book, *Final Quantum Revelation*, he lists nine reasons in support of his hypothesis. They include:

- The 11-year sunspot cycle peaked at the end of June 1908. This sunspot activity manifests itself in Earth's atmosphere by intensified geomagnetic activity and the presence of abnormal optical events such as bright night skies.
- Ball lightning explodes because of the leakage of electric charges from the sphere and the resulting disintegration of its structureless nuclear component.
- Ball lightning typically disintegrates into smaller spheres, which further disintegrate into still smaller spheres, until finally they explode. Eyewitness accounts of many explosions and simultaneous fire

breakouts in widely scattered areas of the forest support a disintegrating ball-lightning scenario.

- Other theories fail to explain the enormous amount of energy accumulated for a short period on the surface of the Tunguska object.

Chukanov also believes that ball lightning can be used to create 'free energy'. The American Physical Society's Bob Park has labelled Chukanov's ideas as 'voodoo science'.

Ghostly geometeors

Andrei Ol'khovatov's geometeor hypothesis is definitely not voodoo science; however, some of his ideas seem ahead of their time. Ol'khovatov, formerly of the Soviet Radio Instrument Industry Research Institute and now an independent researcher based in Moscow, is a popular figure in the large Tunguska cyberspace community, as well as in the small but real community of Tunguska scientific researchers. His website and discussion forum keep the Tunguska debate alive and up to date.

Ol'khovatov became interested in the Tunguska event when in the late 1980s he read about earthquake lights, a glow that sometimes occurs before a large earthquake. He immediately associated these lights with the Tunguska eyewitness accounts. The similarity between accounts of earthquake lights and those of Tunguska led him to believe that there might be some link between the two. He first published his hypothesis in the journal of the Izvestia Academy of Science of the USSR in 1991, but an updated English version appears on his website and in the proceedings of many scientific conferences attended by him. Like all those who reject a cosmic impact, he wants to know where the remnants are. 'Nowhere', he says, 'nothing after decades of detailed research'.

He believes that the explosion was caused by a strong coupling between some unknown subterranean and atmospheric processes. This coupling formed meteor-like luminous objects but of terrestrial origin. For want of a better word he calls these objects 'geophysical meteors' or 'geometeors'. A geometeor resembles a high-speed ball lightning. 'Similar events occur in association with earthquakes (earthquake lights) and in association with a thunderstorm (ball lightning)', he says.

According to Ol'khovatov, the Tunguska region is right in the middle of an ancient volcanic crater. There are many other prominent geological faults, circular structures and other geological formations in the region. Several tectonic faults intersect near the Tunguska explosion centre. There is evidence of increased seismic activity in the region before the explosion. Simultaneously, there was also an increase in anomalous meteorological phenomena: increased sunspot activity; strong increase in thunderstorms; the change in Tunguska region meteorological stations' forecast for 30 June 1908 from 'good weather' to 'bad weather' because of the possibility of a cyclone; and a strong upsurge in atmospheric pressure soon before the event.

This rare combination of large-scale geophysical and meteorological disturbances manifested itself as follows. First, there was luminous activity in the atmosphere in southern Siberia, which was like falling meteors. At about the same time, a swarm of shallow earthquakes started, which were accompanied by brontides (thunder-like sounds of short duration believed to be of seismic origin). Then, at the vent of the crater, there was a large geometeor explosion.

Ol'khovatov believes science is not yet ready to explain the exact mechanism of geophysical and meteorological interactions. However, in his paper he describes in detail how a geometeor can explain various eyewitness accounts and the natural phenomena associated with the event. Ol'khovatov quotes Russian researchers who in 1988 analysed eyewitness accounts of the Tunguska object and found the following descriptions for the shape of the object:

The shape of the Tunguska object was like a …	Percentage of eyewitness accounts
ball	18.8
cylinder	16.3
cone	2.1
star	3.4
tail	14.0
snake	2.3
lightning	2.1
strip of light	2.5
pillar of fire	4.9
flame	10.3
sparks	11.2
other shapes	12.1

He points out that these descriptions hardly conform to a meteorite fall. He also explains the 'three trajectories of the Tunguska meteorite fall' drawn up by researchers from eyewitness accounts by saying that all three trajectories are above the main tectonic faults and they intersect at a point near Vanavara. Does it mean that witnesses saw more than one luminous phenomenon or geometeor? Ol'khovatov says: 'I'm inclined to think that there were several low-altitude fireballs, and that's why there are no reports of two or more fireballs seen simultaneously. The low altitude of fireballs explains why nobody in Vanavara saw a fireball or its trail. Besides the fireballs, evidently there were other typical earthquake lights.'

The three trajectories, according to Ol'khovatov, vary from south-southeast to east-northeast, that is, by up to 90 degrees. These trajectories were drawn from the accounts of eyewitnesses who were 500 kilometres away from the epicentre. 'If it were a meteorite most of the witnesses from the west of the trajectory would say that it flew from the east, while most of the witnesses from the east of the trajectory would say that it flew from the west', he says. 'There would be some witnesses who would say it flew overhead. So, we should have a well-defined trajectory. But there was no such situation in Tunguska. The advocates of the asteroid/meteorite impact choose just a small part of the eyewitness accounts and then announce other eyewitness accounts as "unreliable". As there are several trajectories, so each trajectory has its "reliable witnesses" (a minority), and its "unreliable witnesses" (a majority). A witness for one trajectory could be "reliable" and "unreliable" for all other trajectories.'

At a qualitative level, says Ol'khovatov, his geometeor hypothesis can explain all known facts about the phenomenon. For example, a few years ago several articles appeared in scientific journals stating that about a day or less before an earthquake, the cloudiness level sharply decreases above tectonic faults in the area that will become the epicentre of the earthquake. 'So I studied data from nine meteorological stations within 1,000 kilometres of the Tunguska epicentre (the closest is 500 kilometres away)', Ol'khovatov says. 'I averaged the data from all stations. And indeed, the average daily cloudiness level shows an extremely deep drop on 29 June 1908!'

Ol'khovatov estimates the diameter of the largest ball to be about 1 kilometre. 'But it is just a guess', he adds. He also stresses that the luminosity of the fireball reported was rather weak, while according to the meteorite theories it must be as bright as the Sun, and much brighter near the epicentre – with no persistent trail. He has not estimated the energy of the fireballs: 'The question is still open: Is energy deposited by a lightning ball inside it, or does it also include energy around it? In ball lightning, what we actually see is a tip of the iceberg. I believe science is not yet ready to give a final answer to this question.'

We shall wait. But for many other explanations for the Tunguska mystery we do not have to wait. Science is ready right now to judge them. Whenever the word 'mysterious' is attached to a natural scientific phenomenon that lacks a beyond-a-reasonable-doubt explanation, it becomes a fertile ground for the imagination of those who love outlandish ideas. There is no shortage of eccentric theories for the explanation of the Tunguska event. Let's open the X-files.

CHAPTER NINE

OPENING THE X-FILES

FBI agents Fox Mulder and Dana Scully investigated their own Tunguska mystery in *The X-Files: Tunguska* (episode 9, season 4), in which Mulder encounters a deadly form of alien life living in extra-terrestrial rocks mined from a giant impact crater in Tunguska. But in the following files, dear reader, you're on your own.

Spaceship Tunguska

In August 1945, Little Boy and Fat Man changed our world forever. Curious names for two atomic bombs that unleashed untold devastation upon humanity. On 6 August, Little Boy almost wiped the city of Hiroshima from the map of Japan. Three days later, Fat Man exploded into history in the Nagasaki sky.

'It was hard to believe what we saw', said Colonel Paul Tibbets, the pilot of the B-29 plane that dropped the atomic bomb on Hiroshima, describing at a press conference what he saw seconds after the bomb had been released. 'Below us, rising rapidly, was a tremendous black cloud ... What had been Hiroshima was going up in a mountain of smoke. First I could see a mushroom of boiling dust – apparently with some debris in it – up to

20,000 feet. The boiling continued three or four minutes as I watched. Then a white cloud plumed upward from the center to some 40,000 feet. An angry dust cloud spread all around the city. There were fires on the fringes of the city, apparently burning as buildings crumbled and the gas mains broke.'

On the ground, Kiyosi Tenimoto, a pastor of the Hiroshima Methodist Church, who was about 4 kilometres from the centre of the explosion, saw a blinding flash of light, like 'a sheet of sun', that cut across the sky. Moments later the flash of light had turned into a gigantic mushroom cloud, now known to everyone as the characteristic signature of an atomic explosion. John Hersey, one of the first Western journalists to record the bomb's immediate aftermath, reported in *The New Yorker* magazine of 31 August 1946 that the survivors described the explosion as 'a noiseless flash of light'. He noted that almost no one in Hiroshima recalled hearing any noise of the bomb, but all saw the vast, blinding glare and felt the wave of heat, which was followed closely by the roar of the explosion and its shock. Hersey's extraordinary article, 'Hiroshima' – published simultaneously as a Penguin book which remains in print – had a profound effect on a world which knew hardly anything about the horrors of the atomic bomb.

As for the bomb's incredible destructive power, the numbers speak for themselves. The air temperature at the point of explosion of the 15-kiloton bomb, 580 metres above the ground, exceeded 1 million degrees Celsius. The temperature on the ground at the centre of the blast rose to 6,000 degrees Celsius. The brilliant orange mushroom cloud climbed to 10 kilometres. As the cloud spread it started fires that damaged more than 70,000 houses and killed 140,000 people. But the death toll reached 200,000 due to radiation sickness. In short, two-thirds of an 18-square-kilometre city of 340,000 people was almost obliterated by one atomic bomb in a few minutes.

The world now knew of the immense destructive power of 'the fireball, the mushroom cloud and the intense heat' of an atomic bomb blast. It did not take some Soviet scientists and science fiction writers long to connect the images of Hiroshima and Nagasaki with the images of Tunguska – the fireball, the heat, the thundering noise, the enormous cloud of dust and the devastated forest.

One of them was Alexander Kazantsev, an engineer who had graduated from Siberia's Tomsk Technological Institute in 1930. He was also a well-known science fiction writer who in 1946 published a story, 'The blast', in *Vokrug Sveta*, a popular Russian magazine of science and adventure, in which he presented the bizarre idea that the Tunguska explosion was caused by a 'cosmic visitor' – an extra-terrestrial spaceship, cylindrical in shape and propelled by nuclear fuel. Because of a malfunction the spaceship plunged out of control through Earth's atmosphere, and within a fraction of a second it and its occupants were vaporised in a blinding flash of light. The ETs had come to collect water from Lake Baikal, 800 kilometres from the Tunguska explosion site. This lake – the world's deepest (1,637 metres) and the seventh largest (34,000 square kilometres) – contains the largest volume of surface freshwater. Apparently, the ETs were from a parched planet and very, very thirsty.

Kazantsev's ET hypothesis was well received by other science and science fiction writers. Over the years, as Tunguska expeditions published their new findings, Kazantsev returned to his story again and again, and embellished it into a working theory to explain the Tunguska object. When Florenskiy's 1958 expedition announced the discovery of magnetite globules containing nickel, cobalt, copper, germanium and other elements in the samples collected from the region of the fall, Kazantsev was quick to explain the presence of these elements. In his 1958 article 'Visitor from the Cosmos' (which became the centrepiece of his 1963 book of the same name), he said that the nickel and cobalt came from the outer shell of the spaceship, while the copper and germanium were from semi-conductors and other electrical instruments on board. These and other elements were vaporised when, at the moment of the explosion, temperatures rose tens of millions of degrees. 'In part these elements fell to the ground as precipitation, with radioactive effects', he maintained.

The Soviet Astronomical Journal panned Kazantsev's book as 'a consistent and conscious deception of the reader, in pursuit of one definite goal: to show that he alone, A.N. Kazantsev, has discovered the true nature of the complex phenomena contrary to all the "conjunctures" of the representatives of official science'. But there were admirers as well. One of them was an aircraft designer, A. Yu. Manotskov, who 'proved' that the Tunguska object was under 'intelligent control'. On

average, a meteorite or comet would enter Earth's atmosphere at a speed of 36,000 to 216,000 kilometres per
hour, whereas the Tunguska object 'braked' its speed to
2,400 kilometres per hour, the speed of a jet aircraft. For a
meteorite or comet to plunge down at this low speed, it
would have needed a mass of 1,000 million tonnes and a
diameter of 1 kilometre. Yet this behemoth had made no
crater and left no fragments. Therefore, the Tunguska
object was a small spaceship that was attempting to
land. Kazantsev gleefully agreed: 'Such a tremendous
meteorite would have certainly covered the whole sky.'
Boris Laipunov, a well-known rocket and space travel
expert, also supported Manotskov's reasoning.

Feliks Zigel of the Moscow Aviation Institute added
more meat to Manotskov's assertion that the Tunguska
object was under 'intelligent control'. Some eyewitness
accounts, taken down long after the event, suggest that
the Tunguska body had twice changed course in flight.
This deliberate 'manoeuvre' to change course before
descent was indeed proof that the Tunguska object was a
spaceship flying from another planet, the good professor
declared. He presented another 'proof': the object flew in
an 'enormous loop', first northward then westward,
before crashing; behaviour that appears to exclude a
natural phenomenon. He said that the spaceship had
followed precisely the re-entry angle of 6.2 degrees to the
horizon, which was within the re-entry corridor (between
5.5 degrees and 7.5 degrees) adopted by astronauts
entering Earth's atmosphere. If the angle is too steep, the
spacecraft burns up; too shallow, and it bounces off the
atmosphere like a stone skipping off water.

In an article in the magazine *Znanyie–Sila* in June 1959, Zigel, who is still remembered as 'the father of Soviet UFOlogy', heaped praise on Kazantsev's hypothesis: 'At the present time, like it or not, A.N. Kazantsev's hypothesis is the only realistic one insofar as it explains the absence of a meteorite crater and the explosion of a cosmic body in the air … It is generally known at times – nay, often – new ideas that proved to be most valuable to science were first expressed not by scientists, but by writers of scientific fantasy.' In an interview with the Soviet news agency TASS, he added: 'The more we know of the Tunguska catastrophe, the more confirmation we find of the fact that the UFO which exploded over the forest in 1908 was an extra-terrestrial probe.'

The newspaper *Pravda* of the time, however, considered UFOs 'flirtations with superstitions and religious impulses manipulated indirectly by the Pentagon'. The reference to the Pentagon probably came from the Roswell incident, one of the most famous UFO 'sightings' in American history.

There's a flying saucer in my backyard

On 8 July 1947, the *Roswell Daily Record* broke the news of a cosmic encounter in New Mexico. The story, headlined 'RAAF CAPTURES FLYING SAUCER ON RANCH IN ROSWELL REGION', was based on a press release issued by Roswell Army Air Field (RAAF). When sheep rancher Mac Brazel was making rounds at a ranch 137 kilometres west of Roswell, he found some wreckage consisting largely of rubber strips, wood sticks, tinfoil, plastic, tape with some strange markings that resembled 'hieroglyphics', and very tough paper. Brazel was struck by the unusual nature of the debris. After a few days he drove into Roswell, where he reported the incident to the Sheriff, who reported it to Major Jesse Marcel, Intelligence Officer at RAAF.

The Army closed off the debris site while the wreckage was being cleared. The officers thought that they had found a flying saucer. They shipped the debris to Air Force General Roger Ramey for examination. But in the meantime Colonel William Blanchard, the Commander at Roswell, issued a press release stating that the wreckage of a flying saucer had been recovered. The news caused a sensation around the world, but it was short-lived. Within hours General Ramey called in the local press and announced that RAAF had mistakenly identified remains of a weather balloon as the wreckage of a flying saucer. The next day, the *Roswell Daily Record*'s banner headline proclaimed: GEN. RAMEY EMPTIES ROSWELL SAUCER.

Another story that was not published in the paper, but that some town folks knew about, came from witnesses who had seen the wreckage. They claimed that they had seen alien 'bodies' nearby, describing them as a little more than a metre tall, with bluish skin coloration and no ears, hair, eyebrows or eyelashes. The Air Force explained these 'aliens' as dummies dropped from high-altitude balloons to study the results of the impact. And that was the end of the excitement.

No one ever talked about this episode, at least until the

publication in 1980 of a book, *The Roswell Incident*, which came to the dramatic conclusion that there had been a cover-up of cosmic proportions. In 1988 another book, *UFO Crash at Roswell*, claimed that the US government had found and removed the remnants of the UFO crew – several little alien bodies. These two books were the genesis of a UFO myth and a conspiracy theory – that the government had conspired to cover up the fact that an alien ship had landed at Roswell. The truth is far less exotic: what actually happened was that people who saw the dummies mistook them for aliens.

The term 'UFOs' (unidentified flying objects) was suggested in the mid-1950s by the US Air Force. The term 'flying saucer' was not considered accurate, since many sightings had very natural explanations, while others did not. UFO fans tend to forget that the 'U' in UFO simply stands for 'unidentified', and does not suggest 'extra-terrestrial'.

The question of the existence of extra-terrestrial life (even the simplest form, such as microscopic organisms) and extra-terrestrial intelligent life (advanced technical civilisations capable of communicating with us) is not related to UFOs. Since the dawn of history, humans have been pondering the question: are we alone?

If we are not alone, then how many intelligent civilisations might exist among the stars? In 1961 the American radio astronomer Frank Drake came up with an ingenious approach – now known as the Drake Equation – to answer this question. At that time he was working at the National Radio Astronomy Observatory in West Virginia. In the early 1960s many scientists ridiculed the idea of extra-terrestrial intelligent life, but to Drake the idea of other intelligent civilisations beyond Earth was a distinct possibility. He even placed a sign on his office door: 'Is there intelligent life on Earth?' He was serious about the search for extra-terrestrial intelligent life, and those who questioned his belief in it were not, in his view, 'intelligent life' on Earth.

In its simplest form, the Drake Equation works as follows. To find out the number (N) of advanced technical civilisations in the Milky Way, we need to know:

- How many stars are born each year in our galaxy (R)
- How many of these stars have planets (p)
- How many of these planets are suitable for life (e)
- On how many planets life actually appears (l)
- On how many planets life evolves to an intelligent form (i)
- On how many planets the intelligent life can communicate to other worlds (c)
- The average life of these advanced civilisations (L)

If we multiply these seven factors, we get the equation:

$N = R.p.e.l.i.c.L$

If we know the values of these factors, we can calculate N. As astronomers do not agree on the exact values, estimates of N vary from one (we are home alone) to many millions (yes, a flying saucer could land in your backyard). These estimates are for our galaxy alone, and there are 125 billion (and still counting) galaxies in the presently observable universe. Mind boggling, indeed!

If there are millions of ET civilisations in the universe, then it is possible that some of them might have visited Earth in the past. Drake, who is still involved in the search for extra-terrestrial intelligence at the SETI Institute, says: 'As strongly as I believe that intelligent life exists elsewhere in the universe, I maintain that UFOs are not extraterrestrial visitors. They are the products of intelligent life *on this planet*.'

Zapped by a laser

Intelligent life on this planet in the form of two Russian science writers has suggested that Tunguska was mistakenly zapped by a laser sent by ETs from a giant planet orbiting the star 61 Cygni, about eleven light-years away from us. In a lengthy article published in the magazine *Zvezda* in 1964, Genrikh Altov and Valentina Zhuravleva said that the violent volcanic eruption of Krakatoa in August 1883 generated strong radio waves. This signal was received eleven years later by Cygnian scientists. But they misread the signal as greetings from a distant civilisation.

Following the ages-old Cygnian custom, the courteous scientists decided to send a return message. As their laser technology was more advanced than the radio technology, they directed a laser beam at our planet. Unfortunately, the well-meaning scientists made another mistake. This time they misjudged the Earth's distance and fired a powerful laser beam that zapped the Tunguska taiga. This 'extra strong' Cygnian message was all Greek to the local Evenki people; they did not have the required technology to read their greeting card from the stars.

How could a volcano generate a radio signal? Altov and Zhuravleva simply said that the volcanic ash, flung high into the atmosphere, disturbed the ionosphere, which could have generated a radio signal. The signal was so strong that it reached far out into space. The writers' choice of the star 61 Cygni was obvious: 61 Cygni is a binary star; in 1964 one of its two stars was the only one known to have an extra-solar planet. Now we know that extra-solar planets are common in our galactic neighbourhood. Now we also know that lasers can be used for interstellar communication.

In 1960 Drake made the first real attempt to listen to ETs. In Project Ozma (named after the queen of *The Wizard of Oz*), he aimed a 26-metre radio telescope at the stars Tau Ceti and Epsilon Eridani, some eleven light years away. For two months he listened for radio signals at 1.5 gigahertz, the frequency emitted by hydrogen gas. He chose this frequency because hydrogen is the most common element in the universe. Of course, the search yielded nothing. Since then, more than 100 powerful radio searches have also failed to make any contact. Why have no intelligent radio signals been picked up? Some scientists say that listening to radio signals or sending them might not be the right way to make contact. They suggest that an interstellar laser might be a better communicator.

The American physicist Charles Townes, who was awarded the 1964 Nobel Prize for his hand in the invention of laser, realised around the time that Drake aimed his telescope at distant stars that ET civilisations could just as easily exploit the optical and infrared portion of the spectrum as the radio portion. Decades passed before laser technology had advanced to the point where powerful lasers capable of sending interstellar messages could be made.

Lasers have two main advantages over the millions of radio channels available for broadcasting: light is easier to focus into a tight beam than radio waves; and it is a better carrier of information. A tight laser beam can be easily focused on a target, and it can transmit a whole encyclopaedia in a second – much better than simply asking 'Is anyone out there?' by a radio wave. But lasers are not as cheap as chips (or radios); they require billions of kilowatts of energy to broadcast for a fraction of a second.

Scientists may one day succeed in sending a laser message to their Cygnian counterparts who mistakenly zapped the beautiful Tunguska taiga. An apology is overdue.

Radioactive Tunguska

In his book *Siberia: The New Frontier* (1969), the American author George St George, who spent most of his childhood in Siberia, describes his account of travels there in the mid-1960s. In the book he also makes a brief reference to Tunguska. An excerpt:

> Some investigators seem to believe that whatever flashed across the taiga was intelligently directed because they feel only this explains the changing

course of its flight. Was it then some sort of interplanetary vehicle in trouble, perhaps intentionally destroyed by its crew? Quite a few serious scientists seem to believe so, including some Soviet ones like Felix Zigel. Every serious UFO investigation organization throughout the world lists the Tunguska explosion in connection with possible interplanetary visitors who presumably have visited and studied our planet.

I remember this matter being discussed in our home in Chita, in Transbaikalia, probably in 1914, by my father and his friend, a doctor who claimed to have visited the site of the Tunguska explosion a few months after it had occurred. The doctor had a detailed diagram showing the zig-zag course of the falling body over some 100 miles (where the tops of trees were sheared off) before the actual explosion. He also said that some unusual glow was observed each night over the epicenter of the explosion for weeks after it had occurred, suggesting some sort of radiation. My father, who was interested in the so-called 'flying saucer' lore even then, was convinced that interplanetary visitors were using some parts of the taiga as their terrestrial base. He drew this conclusion from Evenk legends ... Unfortunately all my father's voluminous notes on the subject were lost in China where he died in a Buddhist monastery in 1928.

St George's account is intriguing because it mentions a visit to the site by his father's doctor friend in 1914. This is one of the few known accounts of a visit by a non-Evenki person before Kulik's first expedition in 1927. You can draw another conclusion from this account: even in the dark days of the Tsarist empire, Russians were as fascinated by UFOs as Americans are today. St George also notes that 'no dangerous radiation has been found there at present, so perhaps Tunguska will in time become a familiar tourist attraction as the Arizona crater is today'. The phrase 'no dangerous radiation' is interesting, as it reflects the preoccupation of many Soviet scientists in the late 1950s and early 60s with the idea that the Tunguska site was awash with radiation. The main proponent of this idea was the geophysicist Aleksei Zolotov.

Zolotov was as enigmatic as the Tunguska event itself. In a 1978 special programme to celebrate the 70th anniversary of the event, Moscow Radio described him as 'another noted investigator'. However, in its 70th-anniversary report on Tunguska, the journal *Nature* saw him in a different light: 'His name turns up unfailingly in any discussion of the problem, and his theories, however bizarre to the scientific establishment, do at least get published ... his own academic background seems obscure, and according to one physicist who worked for many years on the Tunguska site, Zolotov was originally simply an oil technologist, co-opted on to an expedition for his knowledge of the local terrain!' In the scientific literature of the time he is referred to as a 'prominent geophysicist'. Ten years later, in 1988, by the time he had led twelve Tunguska expeditions, *Nature* agreed that Zolotov had 'gradually emerged as an authority in his own right'.

One area of Zolotov's authority was his atomic theory of the Tunguska explosion. He garnished Kazantsev's glass of 'exploding spaceship' vodka with a twist of lemon – the explosion was not an accident. In 1975, when he was head of a Soviet scientific team studying the phenomenon, he suggested that the aliens deliberately detonated the spaceship simply to let us know of their existence. The actual area of destruction was 'an amazing demonstration of pinpoint accuracy and humanitarianism', he pointed out.

In 1980, the American science writer T.R. LeMair expanded upon Zolotov's 'humanitarianism' idea in his book *Stones from the Stars*. He claimed that the time of the Tunguska explosion seemed 'too fortuitous for an accident'. If the Siberian missile had met Earth just 4 hours and 47 minutes later, it would have scored a bull's-eye hit on the seat of the tsarist empire; and a tiny change of course would have devastated populated areas of China and India. He suggested that 'the flaming object was being expertly navigated' using Lake Baikal as a reference point: 'The body approached from the south, but when about 140 miles from the explosion point, while over Kezhma, it abruptly changed course to the east. Two hundred and fifty miles later, while above Preobrazhenka, it reversed its heading toward the west. It exploded above the taiga.' A thorough scientific review of eyewitness accounts suggests otherwise: the object did not change its course as it moved across the sky from south-southeast to north-northwest.

Zolotov's major contribution to the Tunguska folklore is not in deciding whether the spaceship exploded by accident or by design, but in the radioactivity it added to the explosion site. He was, in fact, simply cloaking Kazantsev's science fiction ideas in scientific respectability. While the images of 'the fireball and the mushroom cloud' of an atomic bomb made Kazantsev see a spaceship soaring in the Tunguska sky, the images of people dying with atomic bomb radiation convinced him that the 'survivors' of Tunguska were also exposed to Hiroshima-like radiation doses. 'It could be nothing other than radioactivity', explains one of the characters in his science fiction novel *Visitor from the Cosmos*, when a man, shortly after examining the blast area, dies in excruciating pain as if from an invisible fire.

Like Kazantsev, Zolotov had also won Zigel's support. In *Znanyie–Sila* magazine in December 1959, Zigel discussed the results of Zolotov's expeditions of the past three years. During these expeditions, among other things, he compared the effects of the ballistic waves caused by the velocity of the Tunguska body in the atmosphere and the blast waves caused by the explosion itself. Zolotov's study of trees – those that had remained standing and on which the traces of the effects of both waves remained – showed that ballistic waves arrived from the west and broke only small branches, whereas the blast from the north broke larger branches. Zigel estimated the velocity of the body in its final stage of flight to be a relatively low 4,300 kilometres per hour; therefore, the explosion was due to the internal energy of the body, not the energy of its motion. He concluded that the blast waves caused most of the devastation.

Zolotov had found trees some 17 kilometres from the

blast centre which had been subjected to heat and started to burn. He ruled out a natural forest fire. He said that to start a fire in a living tree, the heat energy must be between 60 and 100 calories per square centimetre. Similarly, to have caused a sensation of burning in eye-witnesses 70 kilometres away in Vanavara, the energy must have been not less than 0.6 calories per square centimetre. He estimated the heat energy of the explosion to be about 3.5 megatons. As the estimates of the total energy of the blast were also within this range, he reasoned, the blast was nuclear.

Evidence soon started appearing to support Zolotov's popular story: mysterious scabs suffered by surviving reindeer (burns from hot ash?), tree rings suggesting enormous growth rate after the blast (normal after wildfires?), high levels of radioactive carbon-14 in the soil and peat collected from the region (not enough to support the idea of a nuclear explosion?), and so on.

Although Hiroshima and Nagasaki showed us the horrors of radioactivity, some radioactivity happens constantly all around us. Small amounts of radioactive atoms are found in the soil we stand on, the food we eat, the water we drink, the air we breathe. This is known as background radiation. Our daily dose of background radiation varies from place to place, but average annual levels typically range from about 1.5 to 3.5 millisieverts (150 to 350 millirems). Eighty per cent of this average comes from natural sources such as indoor radon, food and drink, and rocks and soil. The remaining 20 per cent comes from artificial radiation sources, primarily X-rays. From the study of cancers in survivors of

Hiroshima and Nagasaki, scientists have estimated that they were instantaneously exposed to thousands of times the average annual background radiation dose, which continued to increase from long-term fallout.

Most expeditions to Tunguska in the late 1950s and early 60s concentrated on finding the effects of radio-activity on the site. After the 1958 expedition, the Soviet Academy of Sciences decided against any new expedition, but to concentrate on the study of rock and soil samples already collected. This decision led to the formation of the Interdisciplinary Independent Tunguska Expedition (IITE, known as KSE in Russian). The KSE was formed in 1958 in the Siberian city of Tomsk, under the leadership of Gennady Plekhanov, a physician as well as an engineer at the Betatron Laboratory of the Tomsk Medical Institute. KSE was, in fact, formed to discount spaceship theories. The founders even jokingly suggested that 'we must find a nozzle from the spaceship'.

Marek Zbik of the University of South Australia has described the first KSE expedition in the *Bulletin of the Polish Academy of Sciences*. The expedition, led by Plekhanov, included medical students who collected information to test the hypothesis about post-radiation illness among the Evenki people. 'No trace of such illness was detected', Zbik writes. 'They also planned to collect bones from the corpses of Evenki people who had died after the catastrophe ... It was not easy to find such corpses because the Evenki people kept the burial sites a secret.' However, students were successful in examining the bones of people who had died during the smallpox epidemic of 1915. The results of these limited investigations did not confirm an increased radioactivity in the bones tested.

Plekhanov also collected 300 soil samples and nearly 100 plants. An analysis of these samples in Tomsk showed that 'in the centre of the catastrophe radioactivity is one and a half to two times higher than 30 or 40 kilometres away from the centre'. Plekhanov refused to speculate on the cause of this radioactivity. In another study, he compared the bright nights seen in parts of Europe and Asia after the Tunguska explosion with those following the high-altitude nuclear tests conducted by the United States at Bikini Atoll in 1958. He found that both explosions were followed by similar atmospheric effects. In short, Plekhanov failed to find any 'spaceship nozzle'. He left the KSE in 1963 but remains active in research on Tunguska.

Kirill P. Florenskiy and Vassilii Fesenkov, two of the main proponents of the comet theory in the 1960s, violently opposed Zolotov's nuclear explosion ideas. 'There are no planets with a highly organized life from which such a ship could descend', Fesenkov told *The New York Times* in 1960. 'This suggestion has now been rejected by most of the Soviet scientific community', the *Times* added.

Florenskiy devoted most of his time on the 1961–62 expedition to trying to disprove Zolotov. He concluded that the radioactivity at the centre of the blast was within the range of fluctuations of the present background radiation, although he agreed that it was somewhat higher at the centre than it was a few kilometres away. Most of the radioactive atoms were concentrated in the upper layers of soil and peat. He suggested that they had been accumulated from global fallout from atomic and hydrogen bomb tests. On the matter of accelerated tree growth in the devastated area, Florenskiy said that the growth was not due to genetic mutation from radiation but was 'only the normal acceleration of second growth after fires had taken place'.

More recently, in 2001 Academician Nikolai Vasilyev said: 'The results of a search for radioactivity in the region of the Tunguska explosion negate a nuclear hypothesis. It should be noted, however, that a search for traces of radioactivity fallout half a century after a nuclear explosion in the atmosphere is a challenging task, especially taking into account contamination from recent atmospheric nuclear tests.'

The Russian scientists Victor Zhuravlev and A.N. Dmitriev have developed a plasmoid hypothesis for the Tunguska body: a sort of bottle filled with plasma and surrounded by a strong magnetic field. This 100,000-tonne plasmoid was ejected from the Sun. Vitalii Bronshten, the major proponent of the modern comet theory, who died in 2004, was highly critical of this attempt 'to disguise a spaceship as a plasma container'. 'This is a typical ad hoc hypothesis', he said. 'We use this example to demonstrate that all ad hoc hypotheses, a great number of which have been proposed to explain the Tunguska event, are useless.'

Zhuravlev disagrees. Three graphs of magnetic activity found in 1959 at the Irkutsk Magnetic and Meteorological Observatory show a magnetic storm that started soon after the Tunguska event and lasted for about four hours. The magnetograms have nothing in common with those caused by meteorites, but have all the distinctive features of the disturbances of the geomagnetic field that are generated by nuclear bombs. The Tunguska object, says Zhuravlev, was 'a cosmic object the composition and structure of which is unknown to astronomers and physicists'.

The images of a nuclear spaceship exploding above the Tunguska taiga remain vivid in many researchers' minds in the 21st century. One of them is Vladimir V. Rubtsov of the Research Institute on Anomalous Phenomena in the Ukraine. His 'ad hoc hypothesis' is the so-called battle model: in 1908 there was an aerospace battle between two alien spaceships, after which one of them survived and flew back to space. 'Perhaps one day in the future it will be possible to deduce a convincing model of the phenomenon directly from facts accumulated', he writes in a newsletter of the Institute.

Another is Yuri Lavbin of the Tunguska Spatial Phenomenon Foundation in Krasnoyarsk, a group of physicists, geologists and mineralogists who have been organising regular expeditions to the explosion site since 1994. Lavbin believes that the explosion was caused by the collision of an extra-terrestrial spaceship with a comet. In the summer of 2004 Lavbin announced that his team had found two strange black metallic blocks near the site. These 50-kilogram blocks, Lavbin claimed, are the remnants of a spaceship. 'Their material recalls an alloy used to make space rockets, while at the beginning of the 20th century only planes made of plywood existed', he said. The meteorite committee of the Russian Academy of Sciences has dismissed Lavbin's claim, saying that in Siberia where oil geologists regularly work 'you can find a heap of fragments of various machines'.

Spaceship Tunguska still soars in the Siberian sky. Did Tesla's death ray fail to zap it?

Tesla's death ray

Who was Tesla? Nikola Tesla was a genius so ahead of his time that his contemporaries failed to understand his ground-breaking inventions. 'If ever an inventor satisfied the romantic requirements of a Jules Verne novel it was Tesla', said a *New York Times* editorial on 9 January 1943, after his death. 'If that abused word "genius" ever was applicable to any man it was to him.'

He was 'a first-class mathematician and physicist whose blueprints were plausible, even though they were far ahead of the technical resources of his day', the *Times* editorial added. He was so much misunderstood as a great scientist that he became the inspiration for the mad scientist in Max Fletcher's *Superman* cartoons of the 1940s.

An inventor of dazzling brilliance who belonged to the 20th century's heroic age of invention, of which Edison was the most distinguished exemplar, he invented and developed AC power, induction motors, dynamos, transformers, condensers, bladeless turbines, mechanical rev counters, automobile speedometers, gas-discharge lamps that were the forerunners of fluorescent lights, radio broadcasting, and hundreds of other things (the number of patents in his name exceeds 700).

An eccentric who preferred science to society, he

became a virtual recluse for the last quarter century of his life. He never married, never developed any close relationship. 'He made everybody keep at a distance greater than three feet', according to the manager of a New York hotel where he had spent the last years of his life in the company of his pet pigeons.

Tesla, as a recent biography claims, was the man who invented the 20th century, although he was almost forgotten after his death in 1943. But not forever.

On the centenary of his birth in 1956 he was honoured by scientists when they named a unit for measuring magnetism (the SI unit of magnetic flux density, the tesla) after him. This 'nominal immortality' has placed him in the company of Ampère, Volta, Ohm, Gilbert, Henry, Faraday and Hertz, great scientists who have all had electromagnetic units named after them.

Five decades after receiving recognition from his peers, he has now also been given 'cyberspace immortality' from adoring fans in countless web pages of biographies, essays, 'my science hero' projects, online museums, discussion groups, and so on. If the number of web pages can be considered a measure of public popularity, Tesla is now catching up with Marconi but is still a long way away from Edison – two contemporary inventors who have become legends. A surf through Tesla web pages gives the impression that the enigmatic inventor has become a cult hero and has found a place in the hearts of the fans of UFOs, free-energy generators, anti-gravity machines and such other alternative science ideas. Numerous web pages are devoted to his death ray, an invention that links Tesla to Tunguska.

Tesla was born at Smiljan, Croatia (then part of Austria-Hungary) on 10 July 1856. His father was a Greek clergyman and orator, and his mother an inventor of home and farm appliances. After graduating from a high school in Caristadt, Croatia, he studied engineering at the University of Graz. In 1884 he emigrated to America. When the 28-year-old arrived in New York he had four cents in his pocket and a few papers in his suitcase which were scribbled with a drawing and some mathematical calculations for an idea for a flying machine. He lived and worked in New York for almost 60 years. When he died in his hotel room on 7 January 1943 he was penniless, but his room was full of scientific papers and plans so revolutionary that some of them are rumoured to be the blueprints for a missile defence system similar to the US Strategic Defense Initiative (popularly known as 'Star Wars' missile defence) of the 1980s.

His 'practical inventions' were limited to the short period from 1886 to 1903. It was the Jules Verne future that engrossed him, according to the *Times* editorial: 'Communicating with Mars, plucking heat units out of the atmosphere to run engines, using the whole earth as an electrical resonator so that a man in China could communicate wirelessly with another in South America, transmitting power through space – it was to such possibilities that he devoted the last forty years of his long life.'

In later years of his life, Tesla was a favourite of newspaper reporters who revelled in recounting his incredible inventions. On his 78th birthday, Tesla told a

New York Times reporter that he had invented a death ray powerful enough to annihilate an army of 10,000 planes and 1 million soldiers instantaneously. The next day, 11 July 1934, the paper ran a story which was headlined in the style of the time:

TESLA, AT 78, BARES NEW 'DEATH-BEAM'
Invention Powerful Enough to Destroy
10,000 Planes 250 Miles Away, He Asserts.
DEFENSIVE WEAPON ONLY
Scientist, in Interview, Tells of Apparatus
That He Says Will Kill Without Trace.

The story referred to Tesla as 'the father of modern methods of generation and distribution of electrical energy', and quoted him as saying that this latest invention of his would make war impossible: 'It will be invisible and will leave no marks behind it beyond evidence of destruction. This death-beam would surround each country like an invisible Chinese Wall, only a million times more impenetrable. It would make every nation impregnable against attack by airplanes or by large invading armies.'

On his 84th birthday, Tesla declared that he stood ready to divulge to the United States government the secret of his 'teleforce', with which aeroplane motors would be melted at a distance of 400 kilometres, so that an invisible wall of defence would be built around the country. He said that this teleforce was based on an entirely new principle of physics that 'no one has ever dreamed about', and would operate through a beam one-hundred-millionth of a square centimetre in diameter. The voltage required to produce this beam would be about 50 million volts, and this enormous voltage would catapult microscopic electrical particles of matter on their mission of defensive destruction, he added.

Tesla probably conceived the idea for his death ray at Wardenclyffe, Long Island, New York, where in 1902 he built a 57-metre tower and laboratories to experiment on radio waves and on transmitting electrical power without wires. The tower's steel shaft ran 36 metres underground, and it was topped with a 55-tonne, 20-metre diameter metal dome. This experimental facility had the financial backing of the legendary investor J. Pierpont Morgan. However, Morgan pulled out of the venture even before construction was complete. The tower was abandoned in 1911 and demolished in 1917. The main building still stands today.

The popular story that Tesla tested his death ray one night in 1908 goes something like this. In 1908, Arctic explorer Robert Peary was making the second attempt to reach the North Pole, and Tesla requested him to look out for unusual activity. On the evening of 30 June, accompanied by his associate George Scherff atop the Wardenclyffe tower, Tesla aimed his death ray towards the Arctic, to a spot west of the Peary expedition. Tesla then scanned the newspapers and sent telegrams to Peary to confirm the effects of his death ray, but heard of nothing unusual in the Arctic. When Tesla heard of the Tunguska explosion, he was thankful no one was killed, and dismantled his death ray machine, feeling it was too dangerous to keep it.

In a letter to *The New York Times* on 21 April 1907, Tesla wrote: 'When I spoke of future warfare I meant that it would be conducted by direct application of electrical waves without the use of aerial engines or other implementation of destruction ... This is not a dream. Even now wireless power plants could be constructed by which any region of the globe might be rendered uninhabitable without subjecting the population of other parts to serious danger or inconvenience.' Though he believed that it was 'perfectly practicable to transmit electrical energy without wires and produce destructive effects at a distance', there is no evidence that Tesla used the Wardenclyffe tower for his experiments on the death ray.

In an interview with authors Walter W. Massie and Charles R. Underhill for their book *Wireless Telegraphy and Telephony* (1908), Tesla explained his vision of the future uses of radio waves:

> [My experiments] will make it possible for a business man in New York to dictate instructions, and have them instantly appear in type at his office in London or elsewhere. He will be able to call up, from his desk, and talk to any telephone subscriber on the globe, without any change whatever in the existing equipment. An inexpensive instrument, not bigger than a watch, will enable its bearer to hear anywhere, on sea or land, music or song, the speech of a political leader, the address of an eminent man of science, or the sermon of an eloquent clergyman, delivered in some other place, however distant. In the same

manner any picture, character, drawing, or print can be transferred from one to another place. Millions of such instruments can be operated from but one plant of this kind.

He added that he considered the transmission of power without wires more important than his work on radio waves, and that his experiments would show it 'on a scale large enough to carry conviction'. He never realised his vision for the Wardenclyffe tower, because of the lack of financial backers. On 17 February 1905, he wrote to Morgan pleading again for help: 'Let me tell you once more. I have perfected the greatest invention of all time – the transmission of electrical energy without wires to any distance, a work which has consumed my life.' There is no reference in his Wardenclyffe tower works to how his death ray would work. The only reference is in a highly technical article, 'The New Art of Projecting Concentrated Non-dispersive Energy through the Natural Media', written in 1937. In this article he described the actual workings of a particle-beam weapon for destroying tanks.

The death ray may have been a plausible dream, but it was not a reality. Tesla never got the opportunity to test his plans. The Tunguska story seems improbable for another reason. Tesla could not have heard about the Tunguska event before 1928, when stories about it appeared in the American newspapers. Also, there is no record of Tesla's request in Peary's accounts of his expedition. The story has simply been conjured up by joining the dots – Tunguska, Tesla, Peary – with 1908. But the dots do not interrelate.

The American writer Oliver Nichelson, whose name pops up in many books and web pages that link Tesla with Tunguska, believes that the idea of a Tesla-directed energy weapon causing the Tunguska explosion was incorporated in a 1994 fictional biography by another writer, and was the subject of a segment on the TV documentary series *Sightings*. 'Given Tesla's general pacifistic nature it is hard to understand why he would carry out a test harmful to both animals and the people who herded the animals even when he was in the grip of financial desperation', he says. 'The answer is that he probably intended no harm, but was aiming for a publicity coup and, literally, missed his target.'

The evidence is only circumstantial, Nichelson agrees, but he still wants to bet on both heads and tails: 'Maybe the atomic bomb size explosion in Siberia near the turn of the century was the result of a meteorite nobody saw fall. Or, perhaps, Nikola Tesla did shake the world in a way that has been kept secret for over 85 years.' You flip the coin.

Scientists believe that 65 million years ago the dinosaurs were also wiped out by a Tunguska-like, but much larger, fireball. Another similar cosmic impact may lead to the extinction of humans. Thus, the mystery of the Tunguska fireball is inextricably linked to the mystery of the death of the dinosaurs.

CHAPTER TEN

A FIREBALL IN THE DINOSAURS' SKY

Sixty-five million years before the Tunguska fireball. A seaway stretching from the Beaufort Sea (part of the Arctic Ocean) to the Gulf of Mexico divides North America. The regions that some day will be called Alberta and Montana border this inland sea. East of the seaway and the coastal plains rise the newly formed Rocky Mountains.

Between these lowlands and highlands are swamps, lakes, rivers and semi-arid plains. Coniferous and broad-leaved evergreen and deciduous trees, ferns and flower-ing shrubs fill the landscape; grasses have yet to evolve. This wide range of environments provides an ideal place for various species of dinosaurs, the largest beasts ever to roam the land. Among all the types of dinosaur inhabit-ing this prehistoric land, one stands out.

It is 12.5 metres long and 4.5 metres high – as long as a tennis court and tall enough to say hello to you through a second-storey window. It has a terrifying, massive 1.5-metre head and several dozen 18-centimetre long teeth, serrated like bread-knife blades. Its two gigantic bird-like feet have three toes each, but its two-fingered hands and arms appear puny. Its hide is not leathery, but scaly and covered with bumps. The hide is definitely not a

uniform dull green; it's difficult to say, but this dinosaur
may be brightly coloured like its closest relative – the
birds.

Its walking pace is about 5 kilometres per hour. You
can stroll alongside it, keeping up with it without diffi-
culty. When running it can achieve a top speed of 30
kilometres per hour. Certainly not as fast as an ostrich or
a horse, but a good speed when you consider it weighs 10
tonnes, as heavy as three big elephants. No wonder it has
been called a 'roadrunner from hell' – a fitting tribute to
its obvious size and power. Arguably, it is the biggest
meat-eating land animal of all time. It is the lion of its
world, but it is too large, too massive, to be an effective
hunter. It travels and hunts in packs, and attacks like
prowling wolves. Its main diet is plant-eating dinosaurs
that are roughly its size or smaller.

It is the most popular dinosaur of all time – the star of
many monster movies. But it is not the aggressive,
bloodthirsty killer of the movies, and despite popular
belief, it is not necessarily the most vicious animal of its
time. It is not a dull-witted loner either; it is social, lives
in a group and associates with others. It also has a family
life of sorts and probably cares for its young.

Our dinosaur belongs to the species *Tyrannosaurus
rex*, which lived between 67 and 65 million years ago in
Canada (Alberta and Saskatchewan) and the United States
(Colorado, Montana, New Mexico and Wyoming). *T.
rex* is one of the hundred or so species of dinosaurs that
lived in all continents 65 million years ago. It was the
time when dinosaurs ruled the planet. Then they
suddenly disappeared. Their disappearance has all the

ingredients of a thriller: unwitting victims, violent and sudden death, and a mysterious killer. The identity of this killer will also help unravel the mystery of Tunguska.

Vanishing life

Extinction is the disappearance of a species; that is, an entire species of animals or plants has died and can never return. When the environment changes, species must adapt to the new environment to survive. Species that adapt survive, others become extinct. Extinction is not an unusual thing; species disappear continually, and new species appear. In the past, most extinctions were caused by changes in climate or physical surroundings, but one of the main causes of extinctions today is human activity such as the destruction of forests. Half the world's species could disappear within a few decades if we do not change our ways.

When the number of extinctions is very large compared to the number that normally occurs, it is called a mass extinction. In mass extinctions there are few survivors and many victims. Mass extinctions have four important features: (1) many types of species go extinct; (2) large numbers of species go extinct; (3) extinction happens throughout the world – on land and sea; and (4) extinction occurs in a geologically short period.

Throughout the history of Earth there have been five big mass extinctions, one each in the Ordovician, Devonian, Permian, Triassic and Cretaceous periods: in the late Ordovician (438 million years ago); in the late Devonian (380 million years ago); at the end of the

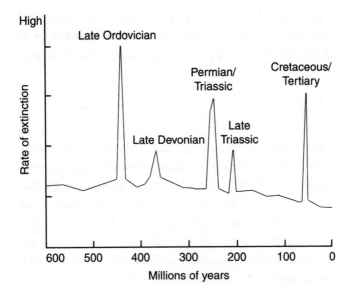

Figure 10: Mass extinctions through time. (Graph based on the work of David Raup and John Sepkoski.)

Permian (245 million years ago); in the late Triassic (208 million years ago); and at the end of the Cretaceous (65 million years ago).

How do we know there were five big mass extinctions? One way of finding out about mass extinctions is to draw a graph of the rate of extinction over time. The peaks in the graph show mass extinctions. In the 1980s, after studying fossils of thousands of species of invertebrates, two American scientists, David Raup and John Sepkoski, drew similar graphs. Their study showed fifteen mass extinctions, of which five clearly towered above the others. Their study also showed a curious pattern: the fifteen mass extinctions seem to be spaced about 26 million years apart. Their conclusion that the mass extinctions have a 26-million-year cycle leads to the question: what causes the cycle? Raup and Sepkoski declared that they favoured 'extraterrestrial causes'. It did not take astronomers long to come up with fanciful ideas, some of which we will soon investigate.

The worst destruction of life in Earth's history took place at the end of the Permian 245 million years ago. Palaeontologists call this extinction the Great Dying because it nearly wiped out most of life on Earth. The death toll included 95 per cent of species in the oceans, 70 per cent of reptiles and amphibians, and 30 per cent of species of insects on land. So many trees and other forms of vegetation disappeared that for a brief period most of the land was covered with fungi. The wiping out of the ruling vertebrates opened the doors for the arrival of the dinosaurs in the Triassic that followed.

What caused this spectacular extinction? The long line-up of suspects includes changes in global climate, sudden drop in sea levels, toxic concentrations of carbon dioxide in the oceans, reduced oxygen and increased carbon dioxide in the atmosphere, decreasing supplies of nutrients in the oceans, and huge volcanic eruptions; but the prime suspect is a massive extra-terrestrial object the size of Mount Everest that slammed into Earth.

The most famous of the Big Five is the one when the dinosaurs died at the end of the Cretaceous. The end of the Cretaceous is often called the Cretaceous-Tertiary, or K-T, boundary (Cretaceous is shortened to 'K' to avoid confusion with the Carboniferous and Cambrian).

Dinosaurs were not the only species to die in the K-T

extinction; possibly 75 per cent of the species living at the time disappeared. Virtually all land and sea animal and plant groups lost species. The main survivors were some land plants, crocodiles, alligators, frogs, salamanders, turtles, birds and mammals. Most of the surviving animals were much smaller than the dinosaurs; they crawled into burrows or hid in water to escape the catastrophe. There was another reason: the surviving animals' place in the food chain. Most of the land animals that died lived in food chains that relied directly on plants. These plants were the first to die during the catastrophe. The surviving animals were in a different food chain. These animals ate insect larvae, worms and other small animals which, in turn, fed on dead and decaying plants.

Fossil records do not tell us whether the K-T extinction was sudden, with everything over in a few minutes, or whether it lasted several million years. The fairest answer is that scientists do not know.

The death star

If mass extinctions have a 26-million-year cycle, then there is a driving force that disturbs the planet at a regular interval of 26 million years. Where is that force in the universe? Where is 'the big clock' that triggers mass extinctions?

It cannot be the sunspot cycle; the number of visible sunspots – the freckle-like, dark, cool regions on the Sun's surface – varies in a regular cycle reaching a maximum about every eleven years. It is not planet Pluto, the most distant of the nine planets, which takes only 247 years to circle the Sun. It is not even the wobble of the Earth's axis, which does not always point exactly at the same spot in the sky but traces out a small circle in a 26,000-year cycle. The 'clock' is certainly not in the solar system. However, the whole solar system makes a complete circle around the centre of the Milky Way in about 250 million years. This is known as the 'cosmic year', but it does not fit the bill either. We do not know of any astronomical event that has a cycle of 26 million years.

The boldest, yet quite feasible, idea to explain the 26-million-year cycle came from the American astronomer Richard Muller and his colleagues. They proposed that, like many other stars, the Sun has a companion star. This companion star moves in an elliptical orbit around the Sun, taking 26 million years to complete one orbit. Once in every 26 million years the companion star comes closer to the solar system, where it passes through the Oort cloud of comets.

During each passage through the Oort cloud, the Sun's companion star disturbs a large number of comets, sending them towards Earth. A shower of comets that lasts thousands of centuries bombards Earth. The increasing dust in the atmosphere darkens the skies. As the temperatures on the ground plunge, most of the animals and plants perish.

The researchers said: 'If and when the companion is found, we suggest it to be named Nemesis, after the Greek goddess who relentlessly persecutes the excessively rich, proud and powerful.' The eminent palaeontologist Stephen Jay Gould proposed that it be named Siva, for the Hindu god who periodically destroys and creates the world.

Popular science magazines were, however, quick to dub it the 'death star'. During the mid-1980s the Nemesis story was very popular and controversial, and was widely covered by the popular press. One newspaper even described it as 'having everything but sex and the Royal Family'. It is rare for a major daily newspaper to write an editorial opinion on a scientific theory. However, in 1985 *The New York Times* wrote an editorial, 'Miscasting the Dinosaur's Horoscope', on the Nemesis theory. It rejected 'the alleged repeating pattern of mass extinctions' and suggested that 'astronomers should leave the astrologers the task of seeking the cause of earthly events in stars'.

The idea of a 26-million-year cycle raises the question: when is the end coming? The good news is: it's at least 13 million years away. The last time Nemesis brushed the comet cloud was about 13 million years ago. It is now about 2 light years (18,921 billion kilometres) away from the Sun, nearly half the distance to the second nearest known star, Alpha Centauri.

If there is a companion star, why have astronomers not yet seen it? Nemesis is believed to be a brown dwarf. Brown dwarfs are too small to achieve the hydrogen burning that powers stars; therefore, they are too faint to detect. No optical or infrared telescope has yet detected a brown dwarf.

In 2001, nearly two decades after the publication of the Nemesis theory, physicists Robert Foot of the University of Melbourne and Zurab Silagadze of the Budker Institute of Nuclear Physics in Russia suggested that Nemesis had proved elusive because it is made of mirror matter and so is invisible. 'It's very hard to prove, but fun to speculate', according to Silagadze.

If it is not Nemesis, the Sun's companion star, behind the 26-million-year cycle, then it must be a distant unknown planet – Planet X. The tenth planet is believed to be three to five times heavier than Earth. It is gaseous like Jupiter, and takes 1,000 years to orbit the Sun. At present it is three times as far out as Pluto, that is, about 15 billion kilometres from the Sun (Earth is only about 150 million kilometres away). Some scientists have proposed that the orbit of Planet X continuously shifts because of the gravitational tug of the other planets. Every 26 million years the shifting orbit disturbs the Oort cloud, producing a comet shower on Earth. No search – even by space probes and the Hubble space telescope – has yet given any hint of the existence of Planet X.

The idea of a 26-million-year cycle in mass extinctions is an attractive one: it suggests that most of them will have similar causes. If we could find the culprit for one, we would have caught a serial killer. Most scientists now reject the idea. However, Raup believes that the idea is still 'alive and well' despite the lack of an astronomical clock. The proposal is still on the table, he says, awaiting new data or new ways of looking at old data. If we move away from the idea of a single cause for all the mass extinctions, we can limit our search to the one that concerns us most – the death of the dinosaurs.

Killer asteroid

Though sedimentary rocks formed during the Cretaceous and Tertiary are limestone, there is a layer of clay at the K-T boundary. Geologists call it the boundary clay. Dinosaurs and other living things that disappeared in the mass extinction 65 million years ago left tell-tale signs of their existence in fossils in this layer of clay. This clay is the site of the mass murder.

Dead bodies – fossils in this case – are not the only evidence of murder. Killers also leave other clues. One day, an American geologist stumbled on a thin layer of clay. The evidence he unearthed opened up a new line of inquiry into the death of the dinosaurs.

In the late 1970s, Walter Alvarez was studying a limestone rock in a gorge outside the northern Italian town of Gubbio. The rock resembled a sandwich. The bottom or older layer consisted of white limestone full of tiny fossils from the Cretaceous. Next there followed a dull red layer of clay about 2 centimetres thick, after which began the top layer consisting of greyish pink limestone, but almost devoid of the Cretaceous fossils. Undoubtedly the clay layer was the boundary clay. Below this layer are the remains of the dinosaurs. Above this layer they are missing.

Some 100,000 tonnes of dust from outer space rain on Earth every year. This invisible cosmic dust is deposited with other sediments when sedimentary rocks are formed. If geologists know the rate at which the cosmic dust falls and how much dust is present in a certain layer of rock, they can find out how long it took to deposit that layer of rock.

The age of the boundary clay can also resolve the question of whether the K-T extinction was a sudden or a slow event that took place over millions of years. 'Sudden' in the geological sense ranges from a few days to hundreds or even thousands of years.

Fossil records show that ammonites, tiny spiral-shelled marine animals, lived right up to the K-T boundary and then disappeared suddenly. It is believed that dinosaurs, who lived at the same time as ammonites and whose fossils are rare, also disappeared suddenly. But fossil records fail to tell us how long it took for the extinction to occur. Alvarez hoped that the Gubbio layer would provide an answer.

When he analysed the clay for cosmic dust, Alvarez failed to find out how long the Gubbio layer took to be deposited. But he discovered something very strange, which provided a crucial first clue to the identity of the mass killer. 'That is what detectives and scientists need: a lot of hard work and occasional lucky break', he remarked.

The lucky break was the discovery of iridium – very large amounts of iridium – in the Gubbio layer. Luis Alvarez, Walter's father, a Nobel-Prize-winning physicist, suggested that the iridium had an extra-terrestrial source. He also predicted that the iridium anomaly, as the whopping amounts of iridium came to be known, should be worldwide. Since this prediction in 1980, Walter Alvarez and other scientists have discovered the same concentrations of iridium as in the Gubbio layer in the K-T boundary clay in Denmark, Spain and New Zealand, and in deep-sea cores from both the Atlantic and the Pacific.

Alvarez, father and son, concluded that some 65 million years ago a large asteroid plunged out of the sky and hit Earth, throwing up a great cloud of dust that quickly covered the planet like a blanket, blocking sunlight for several years. The dust cloud slowly deposited its iridium-rich debris worldwide. This extra-terrestrial impact wiped out the dinosaurs, along with nearly 75 per cent of all other species.

If an asteroid did hit Earth 65 million years ago, where is the impact crater? In 1980, an oil company drilling off the coast of Mexico's Yucatan peninsula stumbled across a great near-circular structure, buried under the surface near the village of Chicxulub, which means 'tail of the devil' in ancient Mayan. It had not been spotted earlier because it was buried beneath 1,100 metres of limestone. No one bothered about the structure until the Canadian geologist Alan Hildebrand learnt about it from a local reporter in 1990. Investigations by Hildebrand and other geologists revealed that the bowl-shaped structure was indeed an impact crater rather than some kind of volcanic structure. It is about 180 kilometres across and 20 times as deep as the Grand Canyon. This estimate of the size is based on boreholes drilled in the search for oil.

The powerful argon-argon technique of dating a rock retrieved from a borehole confirmed that the crater was formed 65 million years ago, when a shallow sea covered the region. Scientists estimate that it was formed by an asteroid about the size of San Francisco zooming at 40 times faster than the speed of sound. The asteroid packed the energy of a 100-million-megaton bomb.

When a giant meteorite or asteroid hits rocks on the

213

Earth's surface, it leaves some evidence of impact. Scientists have found evidence for tektites, shocked quartz and iridium associated with the Chicxulub crater. The K-T boundary layer of the rocks also has large concentrations of iridium, as suggested for an extra-terrestrial impact by Walter Alvarez. Alvarez, who has dubbed the Chicxulub crater 'the crater of doom', believes that it is the best evidence for his theory: 'It looks to me like this is the smoking gun.'

Two other smoking guns have also been found. In 1996 the American scientist Frank Kyte claimed that he had found a pebble in mid-ocean 900 kilometres due west of the Chicxulub crater. The coarse-grain pebble, which is about 2.5 millimetres long, contains iron and iridium in quantities similar to meteorites. The pebble was found in a layer of rock deposited at the same time as the dinosaurs disappeared. Kyte believes it came from outer space: 'There was no way a rock a few millimetres across could get there other than falling from the sky.'

In 1996, after sifting through rocks in the Chicxulub crater, the American scientist Benjamin Schuraytz found two nuggets of iridium. They weigh a few trillionths of a gram and are 99 per cent pure. Schuraytz believes the impact was so powerful that it vaporised other metals, leaving almost pure iridium, which vaporises at more than 4,400 degrees Celsius.

The doomsday

What happened after an asteroid slammed into Earth? There is no shortage of frightening scenarios on how an impact would turn Earth's climate into a killer.

Cold, wet and windy

Within 45 minutes of the impact, a vapour-rich plume of debris would envelop the Earth. There would be enough dust – consisting of roughly equal parts of materials from the asteroid and the Earth's crust – to cause darkness around the world. Scientists have estimated that the Chicxulub impact would have injected 50,000 cubic kilometres of dust in the atmosphere, which settled to form a layer averaging 3 millimetres thick. Without sunlight, photosynthesis would stop. Food chains everywhere would collapse. The darkness would also produce extremely cold temperatures. Scientists call this condition 'impact winter'.

An impact winter can be compared with the scenarios of a 'nuclear winter'. According to these, the explosion of a large number of nuclear missiles would throw huge amounts of dust and smoke into the upper atmosphere, where it would stay for long periods. The result would be darkened skies and lower temperatures for months. In these scenarios, nuclear radiation plays a small part; much more important are the injection of dust and smoke into the upper atmosphere and the subsequent effects on climate.

After a detailed study of plant fossils from 65 million years ago in what is now Wyoming, the American botanist Jack Wolfe has concluded that there was definitely a sudden mass freezing. Fossils all show that the plants are shrivelled, suggesting frost damage. His studies of past climates have even led him to predict when the asteroid hit Earth – on a day in early June. 'Hogwash', says fellow botanist Leo Hickey of Wolfe's findings. Hickey's verdict is based on the study of thousands of leaves from rock layers before and after the dinosaurs' death.

Some experts say that the long winter did not kill the dinosaurs. The discovery of dinosaur fossils in Alaska and southeastern Australia (which was closer to the South Pole 65 million years ago) suggests that dinosaurs could survive many weeks of total darkness. If this is correct, it challenges the argument that darkness and cold caused by an asteroid impact wiped out the dinosaurs.

A massive impact could start earthquakes hundreds of times bigger than the largest one recorded. If the asteroid hit an ocean it would produce immense tsunamis, fast-moving waves higher than skyscrapers which retain their destructive energy while travelling enormous distances. These tsunamis would drown all land areas except mountain ranges. The poor dinosaurs never learned to swim.

The impact might produce winds reaching 1,080 kilometres per hour, according to the American climatologist Kerry Emanuel. These winds could throw huge amounts of dust into the upper atmosphere, changing the climate and destroying the ozone layer. Like many other scenarios, Emanuel's is based on a computer model. Emanuel believes that the Chicxulub impact could easily have caused a catastrophic storm. The immense amount of energy released in the collision would have made the crater extremely hot. Sea water rushing back to cover the new crater would have been heated in turn and would have driven the formation of storms. Emanuel calls these storms 'hypercanes' because they would cause much more damage than normal hurricanes.

Hot, fiery and pungent

If the asteroid hit a limestone rock it would vaporise it. On heating, limestone produces carbon dioxide. As a result the atmosphere would be filled with massive quantities of this gas. The carbon dioxide would trap heat, creating a greenhouse effect with lethal temperatures. To test this theory, two American scientists, John O'Keefe and Thomas Ahrens, shot steel balls from a cannon into limestone rocks at 7,200 kilometres per hour and measured the amount of carbon dioxide released by the impact. They calculated that if a 10-kilometre asteroid were to slam into limestone, it would double the amount of carbon dioxide in the whole atmosphere overnight.

Testing samples of the K-T boundary clay from around the world, in 1985 the American chemist Wendy Wolbach and her colleagues found high levels of carbon, like the soot in the flame of a candle. She believes that globally 70 billion tonnes of soot – the ash of the dinosaurs' world – came from wildfires that started after the impact. The force of the impact created an enormous fireball that spread out, igniting forest fires from North America to Asia. The resulting winds dispersed the soot worldwide, which absorbed the sunlight and thus blocked plant photosynthesis.

Wildfires also created toxic gases such as carbon dioxide, carbon monoxide and methane, which harmed most land life. One estimate shows that the amount of carbon dioxide (10,000 billion tonnes), carbon monoxide

(100 billion tonnes) and methane (100 billion tonnes) released by wildfires was equivalent to 3,000 years of modern fossil-fuel burning. The fires died in the long winter after the firestorms.

Rocks around the Chicxulub crater contain large quantities of sulphur. This has led some scientists to theorise that the blast vaporised the sulphur and spewed more than 90 billion tonnes into the air, where it mixed with moisture to form tiny drops of sulphuric acid. These drops covered the planet like a blanket, blocking sunlight. The blanket remained for decades, pushing the temperatures to near freezing.

Post-impact environment

Scenarios prepared by the American scientists David Kring and Daniel Durda in 2003 show that the post-impact world 'looked, smelled and even sounded different'. Life's diversity saved it from complete extinction, but the new environment was less diverse. Within a year, ferns and algae recovered. After 50 years, shrubs took advantage of the vacant landscape and began to cover it. Trees also began to recover. Re-growth took at least 100 years. Some scientists argue that the process was, in fact, far slower, taking thousands of years; and it took millions of years for life in the oceans to return to normal. 'The impact opened ecological niches for mammalian evolution, which eventually led to the development of our own species', Kring and Durda write in *Scientific American*. 'In this sense, the Chicxulub crater is the crucible of human evolution.'

Those who do not believe in the idea of an asteroid

causing the demise of the dinosaurs argue that the effects of the impact would be limited to a small region only, and could not cause worldwide devastation.

Plumes of steam and smoke

Volcanoes are holes or cracks in the Earth's outer layer from which molten rock or magma – a mixture of liquid lava, solid materials and gases – escapes. Volcanoes are fickle. They can erupt with no warning. Sometimes they just ooze lava. This erratic behaviour makes life difficult for volcanologists, and it probably made life difficult for the dinosaurs 65 million years ago. A volcano suddenly became violent and started spewing plumes of steam and smoke. The sky became dusty and foul, and the fabled beasts choked to death.

Now the argument that a catastrophe wiped out the dinosaurs in the space of a few months or years has three different sides:

- In the vanilla corner are scientists who believe that a rogue rock from outer space did for the dinosaurs.
- In the chocolate corner are scientists who believe that a volcano did the job.
- In the combination vanilla–chocolate corner are scientists who believe in a double whammy – shock waves from the impact of the extra-terrestrial rock immediately spread out beneath the planet's surface, triggering volcanic eruptions on the other side of the planet.

219

Not all explanations are that simple. 'The trail is littered with dead bodies, but there are few clues to how and why the victims died', laments the American geologist Charles Officer, one of the main supporters of the volcano theory. He argues that there were many active volcanoes 65 million years ago. More than a million cubic kilometres of lava erupted in little more than a few thousand years.

The cornerstone of the extra-terrestrial impact theory is the presence of large amounts of iridium in certain rocks. It is presumed that the iridium came from an asteroid or meteorite. But airborne particles from volcanic eruptions also contain large amounts of iridium, so high levels of iridium in impact craters are not evidence of an asteroid hitting Earth.

Gases from a volcanic eruption can also cause acid rain (from sulphur dioxide), greenhouse warming (from extra carbon dioxide) and depletion of the ozone layer (from chlorine). Sixty-five million years ago these effects would have happened on a much larger scale.

'Although it is difficult to see how an impact in, say, China, could kill and dry trees in Europe, recent major volcanic eruptions have shown that a single event can disrupt the world's climate zone', Officer says. He also cites the example of the eruption of a volcano in Indonesia in 1815, which injected so much sulphur dioxide into the upper atmosphere that it led to a worldwide cooling of the atmosphere in the following year. There were cold spells, frosts and crop failures in New England, and the period is still referred to as 'the year without summer'.

Scientists agree that the consequences of an asteroid impact and a massive volcano would be quite similar. The first effect would have been large amounts of dust (either from the crater or volcanic ash), which would darken the skies. The second effect would have been acid rains. These can be produced by nitric acid from chemical reactions caused by the impact, or by sulphuric acid produced by volcanic eruptions.

The idea of 'a second jolt from volcanoes' after the asteroid hit comes from the American geologist Jon Hagstrum. When the asteroid hit Chicxulub, the Earth acted like a giant mirror and the shock waves were focused at an area opposite the impact point. Taking continental drift into account, he estimates that what is India now was 1,600 kilometres or more away from where the focus point was 65 million years ago. There was indeed extensive volcanic activity in India at that time. The volcanoes produced huge lava flows which formed a series of plateau-like giant steps, known as the Deccan Traps (meaning southern stairs). The lava flow extends well over 10,000 square kilometres. It is estimated that the volume of lava is about a million cubic kilometres. Dinosaur eggs and pieces of bones and teeth have also been found in the Deccan Traps.

Died of other causes

Some scientists believe that dinosaurs were not 'killed'; their end was natural. The traditional favourite for the death of the dinosaurs has been the slow change in climate. The list of suspects for a slow climate change includes ice ages, collision of continents and the greenhouse effect. Dinosaurs and many other species were unable to adapt to the changes and therefore died. Mammals and some other species adapted and survived.

Another traditional favourite is the drop in sea level. Large sea level changes are caused either by the movements of the Earth's crust or by changers in the ice caps. Scientists agree that the sea level dropped by 100 metres at the end of the Cretaceous, which caused severe environmental changes. The species died out because of these changes, and the asteroid impact in Chicxulub killed off a few stragglers.

Besides these traditional favourites, there is a plethora of exotic theories. Some samples:

Tiny brains
The dinosaurs died because their bodies continued to grow bigger while their brains remained small. As the dinosaurs became progressively less intelligent, they lost the ability to adapt and survive in a changing environment. This theory was very popular for many decades because those who proposed it could point to the dinosaurs' tiny heads relative to their body size. Yes, we know dinosaurs' brains were not big enough to solve an algebra equation, but the tyrant lizards were smart enough to rule the planet for more than 140 million years. It certainly beats the humans.

Born losers
Biologist David Archibald believes that the dinosaurs died because of bad genes. They were probably born

losers. 'Survival is a game of luck and skill – some species make it, others don't', he says. 'Extinction may always have been on the cards for dinosaurs.' He argues that there is no record of dinosaur extinction throughout the world. No one can say whether dinosaurs died out over-night around the world, or whether they lived for several million years in some places after disappearing elsewhere.

Victims of cancer

A novel but serious attempt to explain the demise of the dinosaurs comes from the American astrophysicist Juan Collar. He claims that dinosaurs were wiped out by epi-demics of cancer. No, it wasn't caused by smoking. The cancer was triggered by massive bursts of neutrinos released by dying stars.

In the final stages of their death, massive stars radiate most of their energy in the form of neutrinos. These dying stars are not as bright as supernovas, and therefore difficult to find. Collar calls them 'silent' dying stars. He predicts that a silent star death occurs within 20 light years of Earth about once every 100 million years. He suggests that a collapsing star would produce twelve malignant cells per kilogram of tissue, each of which could trigger a tumour. The effect would be more severe in dinosaurs because they had more tissue to become cancerous. He advises cataloguing possible 'neutrino bombs' – sources of neutrinos – in the galaxy to save us from the same fate as that of the dinosaurs.

Gamma-ray bath

A new theory is that a gamma-ray burster may have

busted the big beasts. When a neutron star is sucked into a black hole it produces massive bursts of energy that are detected millions of light years away as gamma rays. Some of these bursts also reach our skies, but they are very faint and last only for a few moments.

Thousands of gamma-ray bursts have been detected so far. They are all immensely far away – half-way across the universe. If a burst happened 3,000 light years away in our galaxy, the gamma rays striking Earth's upper atmosphere would create a blue patch in the sky glowing about as bright as the Moon. Much worse events would follow, warn the American astronomers Peter Leonard and Jerry Bonnell. The blast of gamma rays would trigger a chemical reaction in Earth's atmosphere which would wipe out the entire ozone layer.

A few days after the gamma-ray burst, Earth would be immersed in a cosmic-ray bath which would last perhaps for a month. 'At this stage', say Leonard and Bonnell, 'Earth turning on its axis could be portrayed as a chicken roasting on a spit'. The catastrophe would kill 'all but the most well-protected or radiation-resistant species'.

The good news is that a dying neutron star in our vicinity can be predicted many million years in advance. To save Earth, Leonard and Bonnell have a plan: use an asteroid as a shield to block the gamma- and cosmic-ray bath.

Cosmic bullets

Cosmic rays continually bombard Earth from all directions. They are particles such as protons and electrons that travel at very high speeds within our galaxy and elsewhere in the universe. Some of the particles have as much energy as a tennis ball moving at 300 kilometres per hour – they travel nearly as fast as light. Scientists divide cosmic rays into two groups: low-energy and high-energy rays. Low-energy rays are produced in supernovas, the giant exploding stars. Scientists are not yet sure of the source of high-energy rays; they think some of them come from neutron stars. Wherever they come from, how could they kill the dinosaurs?

Two theoretical physicists, John Ellis and David Schramm, seem to have the answer. They say that if a supernova occurred within 33 light years of Earth, it would bombard the upper atmosphere with about 100 times the normal amount of cosmic rays. Such a high radiation would totally destroy the ozone layer, and then the ultraviolet radiation would destroy anything on the surface or close to the surface of the sea. The researchers estimate that one supernova occurs within killing range of Earth once every 240 million years. When a supernova explodes, it ejects enormous amounts of matter which contains unusual isotopes of common elements rarely found on our planet. Earth could have swept up some of this material. The researchers say that these tell-tale signatures of a supernova can be found in the rocks formed at the time.

Cataracts

Dinosaurs were blinded by cataracts caused by excessive ultraviolet light. This was the conclusion drawn by bio-chemist R. Croft in a little book published in the 1980s. He presents a convincing case, but his theory has been challenged on the ground that he had a poor under-standing of dinosaur anatomy. But considering the increased ultraviolet light we receive these days because of the thinning ozone layer, Croft's theory does make sense to us.

The butterfly effect

Chaos, the study of disorder, is an exciting area of science. The weather is the most familiar example of a chaotic – disorderly – system. In a chaotic system, the finest change can bring about a major upheaval. This rule has a flashy title, the 'butterfly effect'. It is possible that an effect as small as a butterfly flapping its wings, say, in Hong Kong can bring about a snowstorm in London.

The American biologist Stuart Kauffman, who has applied chaos theory to mass extinctions, says: 'Mass extinctions, chaos theory suggests, do not require comets or volcanoes to trigger them.' Dinosaurs were part of a chaotic living system, and in such a system superior fitness does not provide a safety net. Their extinction probably occurred for no obvious reason. Explosions and extinctions of various life forms is a pattern that can be found in any chaotic system. 'We are all part of the same pageant', says Kauffman. *Homo sapiens* is not exempt from a fate like the dinosaurs'. Watch out for that butterfly in the garden. Its flapping wings may start the next ice age.

Magnetic reversals

A magnet wiped out the dinosaurs: the biggest magnet on Earth – Earth itself. Earth has a very strong magnetic field which extends some 60,000 kilometres out in space. This tear-shaped magnetic field shields Earth from deadly radiation such as cosmic rays. You can imagine this magnetic field as a big bar magnet inside Earth. It has north and south poles, and is slowly moving. At present, Earth's geographical north and south poles are not pointing in the same direction as its magnetic poles. There is a difference of about 11 degrees.

When certain rocks are formed, small grains of iron act like tiny compasses and line up in the direction of Earth's magnetic field. When the rocks solidify, these little 'compasses' are locked in. The magnetic field is 'fossilised' in the rocks. The study of these rocks shows that the magnetic field has reversed itself many times in the past. Magnetic reversals do not have a regular cycle. Over the past 170 million years, it has reversed nearly 30 times. The last reversal was about 700,000 years ago, when compass needles would have pointed south. No reversal took place for 35 million years during the Cretaceous.

Everything under the Sun has been accused of killing the tyrant lizards. No wonder 'magnetic reversals', as geologists prefer to call the changes in Earth's magnetic field, also appear on this list. The reason? Experts blame these 'reversals' for the ice ages. Earth's magnetic field is slowly weakening. If it continues to weaken at the same rate, the field will completely vanish in a mere 1,500 years – about when the next ice age is expected.

Hot Weather = Stress = Infertility
In 1978, palaeontologist Dewey McLean suggested that

the dinosaurs died because of a slight but critical increase in the global temperature. The effect of the heat was not actually to kill the dinosaurs but effectively to castrate them. Because large animals do not shed excess heat as efficiently as small animals do, a temperature increase of just 2 degrees could have baked the considerable reproductive apparatus of a 10-tonne male dinosaur enough to kill its sperm.

The argument that McLean presented to support his theory went something like this. A large number of completely unhatched dinosaur eggs have been found in rocks, possibly suggesting failure of fertilisation. At the same time, eggs show thinning of the shells. Modern birds under stress also lay thin eggs. Put two and two together and you have a theory: hot weather stressed dinosaurs; stress made them infertile.

Doped

Don't blame the heat – it was drugs. Flowering plants, the angiosperms, evolved around the same time as the dinosaurs died. Many of these plants contain poisonous substances. Modern animals avoid them today because of their bitter taste. Ronald Siegel, an American psycho-pharmacologist, has suggested that dinosaurs had neither the taste for the bitterness nor livers effective enough to detoxify the substance. They died of massive overdoses.

The British palaeontologist Anthony Hallam has looked at the emerging angiosperms from a different angle: the dinosaurs died because of constipation caused by eating the flowering plants that replaced ferns, a dinosaur dietary staple containing laxative oils.

These lateral thinking exercises – poison and consti-pation – have been knocked on the head by those who say that the angiosperms appeared 40 million years before the dinosaurs' death.

More ingenious theories

Self-destruction. The dinosaurs 'self-destructed' them-selves: vicious meat-eating dinosaurs ate all the plant-eating dinosaurs.

Overpopulation. Some blame overpopulation for the dinosaurs' demise. Overcrowding made the females stressed, and stress can cause an imbalance in the hor-mones. Therefore, they laid dangerously thin eggs. Those who propose this theory are certainly walking on eggshells.

Senility. They became senile and forgot to breed and find food. Abnormal and useless features such as the wild neck frills of horned dinosaurs and the bizarre crests of duck-billed dinosaurs prove that the dinosaurs were becoming senile. These features, in fact, enabled dino-saurs to adapt to the changing environment.

Laziness. They starved to death. As their bodies contin-ued to grow bigger, they were no longer able to support their big size with enough food. Fossil evidence contra-dicts this idea. Dinosaurs were not lazybones. They roamed in herds for hundreds of kilometres for food.

Thieving mammals. Clever little mammals developed a

taste for dinosaurs' eggs, raiding their nests to steal them. The dinosaurs could not fight back because the little thieves were too fast for them. Ben Sloan, an American palaeontologist, has a different theory on mammals. He says that some 65 million years ago a receding sea level created a land bridge between North America and the long-isolated Asian continent. Asian mammals invaded North America and began eating the same plants that most dinosaurs ate. 'The mammals ate much less food per animal. But there were so many of them. They ate the last of the dinosaurs out of house and home', he says.

Slipped discs. The dinosaurs suffered from slipped discs, which left them unable to forage for food. A theory without rhyme or reason.

Fussy eaters. They were fussy eaters. 'If they ate mainly one plant, just as the koala bear lives on eucalyptus, they would be in trouble if that plant were no longer available', says James Hopson, an American dinosaur expert.

Itching eyes. 'Itching eyes and dinosaur demise' is the title of a paper written by a distinguished geologist, R.H. Dott Jr. His theory: it was pollen in the air that killed the dinosaurs. A theory not to be sneezed at.

Too much gas! The dinosaurs were wiped out by a serious flatulence problem: the methane expelled by dinosaurs was enough to blast a hole in the ozone layer. The ozone hole in turn damaged vegetation and caused a food shortage which ended the dinosaurs' long reign.

Radiation. In 1984, *Moscow News* reported that dinosaur fossils showed an unusually high uranium content. The report suggests that they may have been killed by radiation in the lagoons in which they lived.

What really happened to the dinosaurs?

The debate on the death of dinosaurs has two sides. On one side are the 'gradualists' who point out that the fossil record shows a steady decline in species diversity starting several hundred thousand years before the end of the Cretaceous. This decline happened because of several environmental changes. The gradualists do not deny an asteroid impact, but they say it only 'killed off a few stragglers'.

'Catastrophists', on the other side, believe that the dinosaurs and 75 per cent of other species were wiped out in the space of few months or a few years. But the catastrophists do not agree on a single cause: some say it was an asteroid; others say it was a volcano; some even believe in 'double whammy' scenarios.

Changes in climate and sea level have occurred throughout Earth's history. These changes take much longer to occur than the extinctions at the end of the Cretaceous. It is possible that these changes played a part in changing the environment, which affected the populations of dinosaurs. Many dinosaur species had been declining before they all finally disappeared: the number of dinosaur types dropped 70 per cent between 73 million and 65 million years ago. It suggests a slow extinction. Extinction is a natural phenomenon, and all species eventually become extinct. Dinosaurs had a good innings. They just ran out of steam.

The supporters of the impact theory would shout: 'No, they were clean bowled by a fireball from the sky.' The American palaeontologist David Jablonski claims that there is widespread agreement in his field that an asteroid or comet did indeed strike Earth 65 million years ago, and generated the huge Chicxulub crater in Mexico. The British palaeontologist Norman MacLeod disagrees: 'Whatever wiped out the dinosaurs was a lot more complicated than a single hammer blow from an asteroid.'

Other exotic theories fail to account for the 75 per cent of other species that also vanished from the face of the planet with the dinosaurs. They all fail the test of a good theory – that it explains as many events as possible. Walter Alvarez says that all the suspects listed under natural causes (from cataracts to mammals eating dinosaurs' eggs) have an airtight alibi: they could not have killed all the different organisms that died with the dinosaurs.

The idea of Nemesis, the Sun's so-called companion star, or Planet X causing mass extinctions every 26 million years is an interesting one. The American palaeontologist Dewey McLean's verdict on Nemesis and Planet X theories is a bit harsh: 'It's science gone absolutely bonkers.' Anyway, most scientists have now rejected the theory.

Walter Alvarez says that murder suspects must typically have means, motive and opportunity. An impact certainly had the means, and the evidence that the impact occurred at exactly the right time points to the opportunity. The asteroid impact theory provides, if not motive, then at least a mechanism behind the crime, he says.

The proof of the pudding is in the eating: a theory can be judged good or bad only when other scientists can test it. The impact theory predicted that all K-T boundary rocks should have whopping amounts of iridium. This prediction was testable, and scientists have found the predicted amounts of iridium in many K-T boundary rocks around the world.

But the cornerstone of the asteroid theory – the large amounts of iridium in certain rocks – has also been challenged. Now it can be proved without any reasonable doubt that the source of iridium can be on Earth (volcanoes) or in outer space (asteroids, meteorites or comets). This leads to 'double whammy' scenarios: an asteroid as well as a volcano. There are some who even go for a 'multiple whammy' scenario: an asteroid, a volcano and a change in climate.

Gerta Keller, a palaeontologist at Princeton University, favours such a scenario. She arrived at this conclusion after studying microfossils at the Chicxulub crater and other sites for more than a decade. Her studies showed that the asteroid struck about 300,000 years before the dinosaurs became extinct, and that the crater was smaller than originally believed. By the time of the impact, there were already many signs of stress in organisms: species were already endangered, their populations having declined and become dwarfed. Instead of an instant 'wipeout', Keller says, this and other mass extinctions can be tied to an intensive period of volcanic activity and resulting greenhouse effects, and probably a series of many asteroid hits. However, she agrees that her theory may not be as riveting as a massive space object hitting Earth. 'Dinosaurs are very popular, and the asteroid theory is sexy, it's a perfect story, and in the past few years it's all you've read in the popular press', she adds.

Everyone agrees that there was a nasty end. Was it sudden or slow? There is no simple answer. Scientists are using the same fossil record, but why are they coming up with different conclusions? The problem is that the fossil record is patchy. Especially when it comes to dinosaurs, the number of known fossils is very small. The American palaeontologist Douglas Erwin has the right advice for his fellow scientists: 'They have to spend more time studying the corpses.'

The debate on the question of the death of the dinosaurs shows no sign of ending soon. Now to another fireball and another debate.

A new fireball in the Siberian sky

Ninety-four years after the Tunguska fireball. About 11.50 p.m., 24 September 2002. A remote, semi-mountainous and sparsely populated region of Siberia near the Vitim River, northeast of Irkutsk and Lake Baikal.

A US Air Force satellite spots an object as it enters the atmosphere, but loses track of it as it falls below 30 kilometres. Moments later a second satellite records a fireball exploding in the clouded sky. A white, bright luminescence appears in the southwest and floods the whole sky. The colour of luminescence changes from white to blue and reddish-brown as it disappears in the northeast.

Virtually no one in Siberia sees the fireball, but local residents hear the sounds of the explosion up to 60 kilometres from the site. The blast waves of the explosion – with the energy of a small atomic bomb – flatten 100 square kilometres of the taiga, but no one is hurt or killed. The explosion also sends seismic waves which make windowpanes rattle and house lights swing in the village of Mama 60 kilometres away. For several days local residents notice sporadic flashing lights in the direction of the explosion.

A mini Tunguska

A small team of researchers and journalists from Irkutsk reached the remote area in late October 2002. But despite using satellite data to locate its position, they were unable to identify and reach the impact site (latitude 58 degrees 9 minutes north, longitude 113 degrees 21 minutes east). The researchers collected 25 eyewitness accounts, which generally agreed that 'a large rock fell from the sky and then the earth trembled'. Some local residents as far away as 70 kilometres from impact site said that they saw a 'sphere with a tail'. Witnesses also 'heard a roar and splashes of light above the taiga far away'.

Alexander Doroshock, a gold miner, said that 'suddenly the sky turned turquoise, there was a large flash followed by an explosion that produced a sharp whistling sound'. Several other witnesses also talked about hearing rustling and buzzing sounds as the fireball streaked across the sky. These were probably electrophonic sounds, which result from the light given off by a meteorite.

Two Mama airport employees, Vera Semenova and Lidiya Berezan, recalled a scary phenomenon: a bright glow at the upper ends of the thin little wooden poles of the fence surrounding the airport's meteorological station. According to Sergei Yazev, the leader of the expedition, the glow was probably caused by a strong electric current produced by the meteorite. As heavy snowfall prevented further searches, the expedition returned to Irkutsk without determining the exact location of the impact site.

In May 2003 an expedition mounted by Kosmopoisk, an organisation of amateur enthusiasts interested in research on various anomalous phenomena, reached the impact site and found an area of about 100 square kilometres covered with fallen trees. Some trees in the centre of this area, where the blast wave touched the ground, showed signs of burning. A few kilometres from this fallen forest they also discovered the impact site, which was covered in about twenty small craters, up to 20 metres in diameter.

On the day after the impact, medical workers in Mama measured radioactivity in local villagers. It showed a two-fold increase above the background radiation, but returned to normal within a few days. Medical workers also told the Kosmopoisk group that the health of local residents worsened for some time after the event. Water and snow in the blast area also tasted bitter. An analysis of water samples in Moscow showed that there were large amounts of tritium – radioactive hydrogen – in the water, 'like in the water of nuclear power station cooling ponds'.

The Kosmopoisk group failed to find any meteorite fragments, and the members believe that the object was probably a small comet with a weight of about 100 tonnes which broke up in mid-air. 'The character of the damage and the radiation background at the explosion epicentre are substantially different from the after-effects of a meteorite fall', says the expedition leader Vadim Chernobrov. However, he adds, the discovery of tritium in water is rather strange for comets. Andrei Ol'khovatov believes that the tritium came from seven underground nuclear tests conducted from 1976 to 1987 about 400 kilometres north of the impact site. 'Siberia is a mysterious place and you could find almost anything you want there, and human activity could make it even more mysterious', says Ol'khovatov.

According to Chernobrov, the members of the expedition do not rule out other rare natural phenomena such as giant ball lightning or ejection of minerals from the ground that disintegrated into water and gas. UFO enthusiasts are not ruling out any cosmic visitor either.

When reporting news of the second largest meteorite, after the Tunguska meteorite, to fall in Siberia within a century, newspapers around the world were more interested in speculating what would have happened if the fireball had exploded over a big city. 'If it hit Central London, Britain would no longer have a capital city', speculated the London *Times*. 'It was an explosion of such a force that if the supposed fireball had fallen on Moscow, half of the Russian capital would have turned into desert, and the other half into ruins', echoed the Russian news agency Novosti. The news of the Vitim fireball also made asteroid doomsayers worried, because such an impact would force them to revise their estimate for the likelihood of a devastating asteroid striking Earth (meaning asteroid armageddon may be closer than you think). You have not yet heard the last word on the Vitim fireball or the Tunguska fireball.

Why must we find the answer?

The Tunguska explosion was a cataclysm that has happened countless times in Earth's history, and it is sure to happen again – that's what Academician Nikolai Vasilyev (1930–2001) believed. 'Had such a cosmic body exploded over Europe instead of the desolate region of Siberia, the number of human victims would have been 500,000 or more, not to mention the ensuing ecological catastrophe', he said. 'The Tunguska episode marks the only event in the history of civilization when Earth has collided with a truly large celestial object, although innumerable such collisions have occurred in the geological past. And many more are bound to occur.'

He stressed that continued investigations of the Tunguska event were important, simply because it would happen again. Only by knowing what the object was, and by knowing its devastating biological consequences, will the scientific and medical communities be in a position to deal with such a cataclysm in the future. 'Today the

Tunguska problem can be considered as an important part of the larger problem of possible collision of Earth with near-Earth asteroids', Vasilyev said. The most important question for us is: can it happen again? We'll never know the answer.

Like the riddle of the death of the dinosaurs, the Tunguska mystery still eludes scientists. Many believe that this cosmic mystery would finally be solved only if we found ponderable fragments of the object. At least, let's try to guess whodunit.

CHAPTER ELEVEN

WHODUNIT?

Not that long ago, a popular Russian website asked its visitors to answer the question: what do you think the Tunguska object was? The results of the poll, which was restricted to one vote per e-mail address, were as follows:

A comet	31%
A meteorite/asteroid	27%
An alien spaceship	9%
Other	33%

The website is frequented by science-oriented Russians, and this is reflected in the low rating of the alien space-ship theory, which is believed to be very popular among Russian Tunguska fans.

It is surprising that, after eight decades, international scientists remain as divided as Russian web surfers. Chris Trayner of Leeds University seems to have solved the riddle of why scientists have failed to solve the riddle of Tunguska. He says that many of their problems stem from the old East–West divide, and poor access to the Russian literature. He suggests that to disentangle what might have happened in 1908, scientists need extensive translations and online indexes of the Russian source material. The same must be said of English sources for Russian scientists.

In the meantime, let's try to disentangle what really happened. There are seven basic sources of information on the event: (1) the devastated forest and its pattern of damage (first observed nineteen years after the event); (2) records of atmospheric and seismic waves at the time of the event; (3) records of magnetic storms at the time of the event; (4) bright nights observed in parts of Europe and Asia after the event; (5) anomalous atmospheric phenomena observed after the event; (6) study of microscopic particles found at the explosion site and in Antarctica; and (7) eyewitness accounts (first collected thirteen years after the event).

Almost everyone agrees on the following points:

- The accurate time of the event (0014 GMT; 7.14 a.m. local time) and the exact location of the epicentre (latitude 60 degrees 55 minutes north, longitude 101 degrees 57 minutes east).

- A large incoming object, presumably a meteorite or a comet, was seen over an area 1,500 kilometres across. The object's brightness was comparable to that of the noonday sun.

- It exploded in mid-air, between 5 and 10 kilometres above the ground. The energy of the blast was probably between 10 and 20 megatons.

- The debris was blown upwards into the stratosphere. No significant fragments, except microscopic globules at the epicentre and in Antarctica, have been found.

- The blast created a shock wave that levelled 2,150 square kilometres of forest, of which about 200 square kilometres had been burnt by a heat wave. In a forest fire, trees are usually burnt on the lower part of their trunks, but these trees had been burnt uniformly and continuously.

- There is no impact crater at the epicentre, the point under the explosion where the shock wave first contacted the ground.

- The devastated forest has the shape of a butterfly spread over the ground. The epicentre lies near the 'head' of the butterfly.

- The heat and shock waves were felt by many eye-witnesses at Vanavara, about 70 kilometres from the explosion site. Bright lights in the sky were seen as far as 700 kilometres away; and loud explosions, like gunfire, were heard after the explosion up to 1,200 kilometres away.

- The blast also created disturbances in the atmosphere which were recorded around the world. The impact of the blast on the ground generated seismic waves which were recorded well beyond Russia.

■ A local magnetic storm – similar to ones produced by nuclear explosions in the atmosphere – began a few minutes after the blast and lasted for about four hours.

■ The explosion caused very bright nights in parts of Europe and Asia which lasted for several nights. Noctilucent clouds were also observed. The dust in the stratosphere caused these optical anomalies.

■ There was accelerated growth in young trees that survived the blast.

The controversial points:

■ The object's shape as it raced across the sky before exploding: a 'pipe', 'pillar' or 'tube', as mentioned by some eyewitnesses? Did it leave a trail of smoke and dust?

■ Its nature, size and mass; for how long it was seen in the sky.

■ Its entry angle, its subsequent flight path, and its velocity immediately before explosion; the duration of the explosion.

■ The flight ending in multiple explosions.

■ Increased radioactivity at the explosion site; genetic effects on the local population; the accelerated growth of trees being caused by genetic mutation.

- Craters such as the Suslov crater being formed by the impact of the exploding object.

- Optical anomalies being observed before the event as early as 23 June.

Bordering on the ridiculous:

- The object twice changed course in flight.

The usual suspects

As far back as 1969, a Russian writer, P.I. Privalova (believed to be the pseudonym of Igor Zotkin, a member of the then Committee of Meteorites of the Soviet Academy of Science), published a list of 77 theories that had been put forward to explain the Tunguska event. The list could be stretched to 120, Ms Privalova hinted, if one was at a campfire in the taiga with a glass of vodka. As the following list has been compiled without the benefit of a glass of vodka at a taiga campfire, it includes only a 'dirty dozen'.

A comet that disintegrated in the atmosphere. The unusually loose structure of the comet led to its break-up in the atmosphere. The diameter of its nucleus is estimated to be 40 metres, much smaller than diameters of visual comets.

An asteroid that exploded in mid-air. A stony asteroid that exploded at about 8 kilometres above the ground. It was about 30 metres across. The 15-megaton explosion released a million tonnes of small particles in Earth's atmosphere. Winds dispersed the dust in the stratosphere, which caused the bright skies reported in the aftermath of the fall.

A mini black hole that passed through Earth. A mini black hole, invisible to the naked eye, entered at Tunguska and then travelled through Earth for about 15 minutes. When it exited through the North Atlantic, it caused shock waves in the ocean and the atmosphere.

An anti-matter rock that annihilated itself when it entered the ordinary matter atmosphere. The explosion, which was like a 35-megaton atomic or hydrogen bomb, generated trillions of radioactive carbon-14 atoms.

A mirror matter rock that nobody could see. As it dived into the atmosphere, the heat caused it to explode at high altitude. The rock was roughly 100 metres in size and weighed about 1 million tonnes.

A volcanic blow-out. Natural gas escaped from narrow underground volcanic vents and raced upwards at high speed and started mixing with air. After a few hours this volatile mixture, which contained 10 million tonnes of methane, exploded like a fireball.

A giant lightning ball that materialised from nowhere. The estimates of its diameter vary from 200 metres to 1 kilometre. It disintegrated into smaller spheres, which further disintegrated into still smaller spheres, until finally they exploded.

'Geometeors' that came from below. The explosion was caused by a strong coupling between some unknown subterranean and atmospheric processes. This coupling formed meteor-like luminous objects, but of terrestrial origin.

A plasmoid surrounded by a strong magnetic field. This 100,000-tonne plasmoid was ejected from the Sun to wreak havoc at Tunguska.

An alien spaceship that broke down. It plunged out of control through Earth's atmosphere and within a fraction of a second it vaporised in a blinding flash of light.

A zap from an alien laser. A laser sent by ETs from a giant planet orbiting the star 61 Cygni zapped the taiga.

An experiment on a 'death ray' which got out of hand. Nikola Tesla mistakenly pointed his death ray at Tunguska.

In the witness box

A tiny comet with a long tail. The available evidence on the object's orbit is consistent with the orbits of the Earth-crossing asteroids, but not with the orbits of short-period comets. The collision of comet Shoemaker-Levy 9 with Jupiter has shown us that the mass of a comet that enters a planet's atmosphere needs to be more than 100 million tonnes in order to trigger a powerful explosion at the end of its journey. The Tunguska fireball's pre-explosion mass is believed to be 1 million tonnes. It is inconceivable that such a small comet could survive the intense atmospheric pressure on its journey to the Siberian sky. On the other hand, an asteroid could survive such an arduous journey. It has to be an asteroid, a stony asteroid.

A stony asteroid. The facts that prove that the colliding object was a comet include its unusually loose structure, which led to its break-up in the atmosphere; its dust tail, pointing away from the Sun, which caused unusual sunsets; and the nature of its orbit. The Sun's glare prevented any sighting of the comet before it hit Earth, because its direction and the angle of strike towards Earth were from behind the Sun. It was not an asteroid for another reason: the absence of reasonably large asteroid fragments. No fragments, no asteroids. It was certainly a comet, a ball of ice and dust.

A mini black hole. It's not theoretically impossible that a mini black hole could pass through Earth, but the black hole theory was a good try by a couple of theoretical physicists. It flopped when no one could find any record of a black hole's exit on the other side of Earth through the North Atlantic. Surely it was a big rock, not an invisible black hole.

An anti-matter rock. The trouble with these rocks is that they can't survive the journey through the atmosphere. Anti-matter–matter annihilation also annihilates this theory.

An invisible mirror matter rock. A good effort to test drive the new mirror matter theories. The effect on Tunguska, however, has been invisible. It wasn't a rough rock, it was a beautiful ball.

A lightning ball that turned into a killer. No one has yet explained how a tiny lightning ball could turn into a giant fireball. But it's easy for a cloud of methane.

A volcanic blow-out. The radial pattern of fallen trees shows that the shock waves came from above, not from the rattling of Earth. What about the fireball that eyewitnesses saw streaking across the sunny Siberian sky? Try a geometeor (but the same arguments can be applied against that theory).

Ghostly geometeors. Meteors that do not come from the sky, but from the ground under your feet? A plasmoid sounds more plausible.

A plasmoid. A sort of bottle filled with plasma and surrounded by a strong magnetic field? A spaceship disguised as a plasma container? Why not try the real thing?

An alien spaceship. Was it on its way to the *Star Wars* studios and got caught in a space-time vortex which spewed it onto the set of *The X-Files: Tunguska*? A laser beam is a better bet.

A sharp laser beam. You are not alone, earthlings. Read your next postcard from space carefully, otherwise you'll be zapped by a death ray.

A deadly death ray. A yarn that seems to have been spun on April Fools' Day. Our benevolent Dr Tesla fired his death ray to deflect the asteroid that was on its way to destroy his home city of New York. The death ray changed the course of the rogue rock and it exploded in almost empty Siberia. Thank you, Dr Tesla.

The experts' testimony

Academician Nikolai Vasilyev: 'Though the final choice between the asteroid and the comet theories has not yet been made, the chances of the stony asteroid version have recently grown substantially … The comet theory has obviously lost its dominating position, but it is not clear whether the pressing problems can be solved within the asteroid alternative.'

Dr Vitalii Bronshten: 'We astronomers know two types of solar system bodies – comets and asteroids. We do not know anything else.' (A remark made to Dr Ol'khovatov about the geometeor theory.)

Dr Andrei Ol'khovatov: 'If for many years you were thinking about a meteorite or an alien spacecraft, of course, it would be very hard to change your point of view.' (This may disqualify you from jury duty.)

Betting on the verdict

In this race of dirty dozen suspects, the favourite in a betting shop – run by astronomers – would be a rogue rock from space: not small enough to burn up completely in the upper atmosphere; not big enough to leave a hole in the ground.

Poor Kulik died believing that the rock was big enough to dig a big crater and leave some visible remnants. But we have now a computer ID of the errant rock: a small stony asteroid 30 metres across, moving about 54,000 kilometres an hour. It exploded 8 kilometres above the Siberian wilderness, leaving only a devastated forest.

However, some astronomers dispute this ID. They say the rock was not an asteroid but a small chunk of a comet. According to Duncan Steel, the chunk was from a large comet that has been disintegrating for tens of thousands of years, and this disintegrating comet would have been a threat even to prehistoric humans. Its impact risk peaked around 5,000 to 5,500 years ago, when the work first began on Stonehenge, the most important of all prehistoric monuments. Steel's calculations suggest that it was originally built as a prehistoric early warning system for impending impacts by Tunguska-like objects. On the day of greatest danger, the comet debris would have been visible just above the horizon in line with the 'Heel Stone' set outside the main circle, he says. A fascinating idea, indeed.

Killer-rocks-from-space scenarios are beloved of astronomers who see Earth-crossing asteroids or comets as a potential threat to humanity. 'The problem with the scientists involved in the research is that they get very emotional and less rational about the science', says Elmar Bartlmae, a German filmmaker who has produced a documentary on the Tunguska event.

What would be the verdict of a dispassionate science writer who has no pet theory to push and who has diligently investigated this cosmic whodunit? Without reading my stars I would wager on a cause that lies not in the stars but beneath ground.

Some scientists have thought about this mystery outside the square (or the elliptical orbit of an asteroid or a comet). Wolfgang Kundt is one of them. After analysing statistics of extraterrestrial and terrestrial events he came to the conclusion that tectonic events such as earthquakes and volcanoes are 20 times more frequent than asteroid or comet impacts at the same destruction energy. He says that the disappearance of Sodom and Gomorrah in biblical times may also be due to a volcanic eruption, making the two cities slide to the bottom of the Dead Sea.

His arguments for a tectonic explanation (pages 154–60) for the Tunguska event – a volcanic blow-out in which millions of tonnes of methane gas escaped from a volcanic vent and turned into a fireball – are perfectly plausible. When the gas was suddenly released it rose, burned and formed clouds in the upper atmosphere for several days, where they scattered sunlight. The scattered sunlight at night gave rise to the bright nights of Tunguska.

'We live on a tectonically active planet', he says. 'To me, the case of Tunguska is crystal clear, multiply redundant.'

There's just one hitch: Kundt's explanation fails to account for the fireball that was seen by many witnesses racing across the sky. Andrei Ol'khovatov, who also favours a terrestrial explanation but a different one from that of Kundt, has analysed witnesses' accounts. He says that these accounts do not conform to those expected from a giant space body. 'Remarkable that, in the area within 200 kilometres around the epicentre, witnesses did not see any bolide or just even its trail, at least, but they saw flames shooting up, and fire pillars, flashes in the sky and other non-bolide luminosities', he says.

Ol'khovatov's so-called geophysical explanation (pages 167–72) combines earthquakes and ball lightning. It has certain merits, but I will bet my proverbial two bob on Kundt's tectonic theory.

TIMELINE: ONE HUNDRED YEARS OF AN ENIGMA

1908 7.14 a.m., 30 June. A fireball explodes in mid-air
near the Stony Tunguska River in Siberia and
flattens a vast forest. Nearest eyewitnesses are 70
kilometres from the explosion site, but villagers as
far as 700 kilometres away see bright lights in the
sky and villagers as far as 1,200 kilometres away
hear loud explosions. Siberian newspapers report
the explosion, but they are not sure of its nature;
some suggest it to be a meteorite. After the explosion, bright 'night glows' are observed in parts of
Europe and Asia. This unusual phenomenon is
widely reported in newspapers in Britain, Europe
and the United States, but no one knows its cause.
An observatory 970 kilometres from the explosion
site records a magnetic storm that began a few
minutes after the blast and lasted for about four
hours. Seismic waves are recorded around the
world. Six microbarographs in England record air
waves created by the blast.

1910 The first expedition to the Tunguska site by a non-
Evenki person. A wealthy Russian merchant and
goldsmith named Suzdalev is rumoured to have
discovered diamonds at the site.

1921 September. Nothing is heard of the explosion until Leonid Kulik, a Russian scientist, is assigned the task of locating and examining meteorites fallen in inhabited regions of Russia before and after the First World War. During his expedition to Siberia, Kulik learns of a meteorite that had fallen near the Stony Tunguska River. The expedition ended without him visiting the explosion site.

1924 S.V. Obruchev, a Soviet geologist, conducts geological studies in the Tunguska region (but not the explosion site).

1925 A.V. Voznesensky, Director of the Irkutsk Magnetic and Meteorological Observatory in 1908, claims that the seismic and air waves recorded by his observatory on 30 June 1908 were both caused by the fall of a giant meteorite.

1926 I.M. Suslov, a Soviet ethnographer, visits Tunguska region. (The famous Suslov crater is named after him.)

1927 21 May. The first Tunguska expedition, led by Kulik, reaches the epicentre of the explosion. Kulik becomes the first scientist to visit the site.

1928 June. Kulik's second expedition reaches the explosion site. Kulik's expeditions are widely reported in British and American newspapers.

1929 Kulik's third expedition. A British scientist notices the coincidence of the date of the Tunguska explosion and the airwaves recorded in England on 30 June 1908.

1930 Other British scientists suggest that the airwaves recorded in England and the remarkable night glows of 1908 were caused by the Tunguska meteorite.

1934 The British scientist F.J.W. Whipple and the Russian scientist I.S. Astapowitsch independently propose that the Tunguska object was a comet. (Astapowitsch, in fact, expanded the idea suggested by Academician Vladimir Vernadsky.)

1938 The first aerial photographic survey of the Tunguska region is undertaken.

1939 Kulik's fourth expedition to Tunguska. The last expedition before the start of the Second World War.

1941 The American meteorite expert Lincoln La Paz suggests that the Tunguska object was a contra-terrene (anti-matter) meteorite.

1942 14 April. Kulik dies in a German prisoner-of-war hospital.

1946 The Russian science-fiction writer Alexander Kazantsev publishes a story suggesting that the Tunguska object was an alien spaceship.

1957 The Russian mineralogist A.A. Yavnel microscopically analyses soil samples brought back by Kulik in 1929 and 1930. These samples were later proved to be of terrestrial origin.

1958 30 June. The Soviet Union releases a commemorative 40-kopeck stamp featuring a portrait of Kulik, to celebrate the 50th anniversary of the Tunguska event. The fifth expedition, the first after the Second World War, led by Kirill P. Florenskiy. The Interdisciplinary Independent Tunguska Expedition (IITE, known as KSE in Russian) is formed in the Siberian city of Tomsk to discount space-ship theories.

1960 Florenskiy supports the comet theory in an article in the US magazine *Sky & Telescope*. Academician Vassilii Fesenkov also presents arguments in favour of the comet theory to *The New York Times*. The Soviet Academy of Sciences builds a simple memorial on Kulik's grave in the town of Spas-Demensk, about 300 kilometres southwest of Moscow.

1961 Florenskiy leads the sixth expedition which continues into 1962.

1964 The Russian science-fiction writers Genrikh Altov and Valentina Zhuravleva suggest that Tunguska was zapped by a laser beam sent by ETs.

1965 The American scientists Willard Libby, Clyde Cowan and C.R. Alturi present a detailed theory showing that the Tunguska object was made of antimatter.

1959 Feliks Zigel, so-called 'Father of Soviet UFOlogy', suggests that the Tunguska object was a UFO.

1966 The English translation of E.L. Krinov's book *Giant Meteorites* is published. Its 141-page section 'The Tunguska Meteorite' has an authoritative account of early research on Tunguska.

1973 The American theoretical physicists A.A. Jackson IV and Michael P. Ryan, Jr say that the Tunguska object was a mini black hole, which passed through Earth and exited through the North Atlantic ocean.

1976 The first Tunguska book in English is published: *The Fire Came By: The Riddle of the Great Siberian Explosion* by John Baxter and Thomas Atkins.

1975 The Israeli scientist Ari Ben-Menahem concludes that the explosion took place 8.5 kilometres above the ground and had energy of about 12.5 megatons of TNT.

1977 The British scientist Anthony Lawton suggests that the Tunguska fireball was in fact a giant lightning ball.

1978 The Slovak astronomer Lubar Krésak suggests that a piece of comet Encke had exploded at Tunguska.

1983 The American scientist Zdenek Sekanina proposes that the explosion was caused by a stony asteroid. The American scientist Richard Turco suggests that the bright nights were caused by noctilucent clouds produced by the dust that reached the stratosphere. The American chemist Ramachandran Ganapathy says that the globules collected by Florenskiy's 1961–62 expedition are enriched in iridium, a metal that is abundant in extra-terrestrial bodies, and contain other evidence of extra-terrestrial origin. He also discovers traces of the Tunguska fireball in an Antarctic ice core.

1984 The Russian scientists Victor Zhuravlev and A.N. Dmitriev present their plasmoid hypothesis.

1989 The first post-Cold War expedition open to international scientists.

1991 The first Italian expedition, led by Menotti Galli and Giuseppe Longo. The expedition collects particles from resin in Tunguska trees. The particles contain some elements which are commonly associated with stony asteroids. The Russian scientist Andrei Ol'khovatov publishes his 'geometeor' theory. Geometeors are meteor-like luminous objects but of terrestrial origin.

1993 American scientists Christopher Chyba, Kevin Zahnle and Paul Thomas give the asteroid theory new weight and rigour. They say that the explosion released about 15 megatons of energy in the atmosphere at an altitude of about 8 kilometres.

1994 An unknown American writer suggests that the explosion was caused by a Nikola Tesla experiment on a death ray which got out of hand.

1996 The Russian scientist Vladimir Svetsov shows that the entire mass of the Tunguska object vaporised before it could reach the ground. Ablation of the Tunguska debris was total.

1998 Sekanina revisits his asteroid theory and presents new arguments in favour of it. The Russian scientist Vladimir Alekseev suggests that the flight of the object ended in multiple explosions which were responsible for gunfire-like sounds heard by eyewitnesses.

2001 Academician Nikolai Vasilyev, who coordinated the scientific research of 29 Tunguska investigations from 1963 to 2001, dies. A team of Italian scientists, based on an idea of the late Paolo Farinella (1953–2000), calculates 886 valid orbits of the object, of which 83 per cent are asteroid orbits and 17 per cent comet orbits. The German astrophysicist Wolfgang Kundt suggests that the explosion was caused by 10 million tonnes of methane gas which escaped from a volcanic vent.

2002 The Australian physicist Robert Foot suggests that

the Tunguska blast was caused by a mirror matter asteroid.

2004 Dr Vitalii Bronshten, a well-known Tunguska researcher and the main supporter of the comet theory, dies.

2008 As we prepare to celebrate the 100th anniversary of the Tunguska event, there is still no final answer to the question: what really caused the explosion?

SOURCES AND FURTHER READING

Books

On Tunguska

John Baxter and Thomas Atkins, *The Fire Came By: The Riddle of the Great Siberian Explosion*, Macdonald and Jane's, London, 1976

Rupert Furneaux, *The Tungus Event*, Nordon Publications, New York, 1977

Roy A. Gallant, *The Day the Sky Split Apart: Investigating a Cosmic Mystery*, Atheneum Books for Children, New York, 1995

E.L. Krinov, *Giant Meteorites*, trans. J.S. Romankiewicz (Part III: *The Tunguska Meteorite*), Pergamon Press, Oxford, 1966

Jack Stoneley, *Cauldron of Hell: Tunguska*, Simon and Schuster, New York, 1977

Others

Walter Alvarez, *T. Rex and the Crater of Doom*, Princeton University Press, Princeton, 1997

David H. Childress, *The Fantastic Inventions of Nikola Tesla*, Adventures Unlimited Press, Illinois, 1993

Frank Drake and Dava Sobel, *Is Anyone Out There? The Scientific Search for Extraterrestrial Intelligence*, Delacorte Press, New York, 1992

Robert Foot, *Shadowlands: Quest for Mirror Matter in the Universe*, uPUBLISH.com, 2002

Peter Goddard (ed.), *Paul Dirac and His Work*, Cambridge University Press, Cambridge, 1998

Richard Muller, *Nemesis: The Death Star*, Weidenfeld and Nicolson, London, 1988

Richard O. Norton, *Rocks from Space*, Mountain Press Publishing, Montana, 1994

Charles Officer and Jake Page, *The Great Dinosaur Extinction Controversy*, Addison-Wesley, Reading, 1996

Albert Perry, *Russia's Rockets and Missiles*, Macmillan, London, 1960

David M. Raup, *The Nemesis Affair: A Story of the Death of Dinosaurs and the Ways of Science*, W.W. Norton, New York, 1997

George St George, *Siberia: The New Frontier*, Hodder and Stoughton, London, 1969

R. Taton and C. Wilson (eds), *The General History of Astronomy*, Cambridge University Press, Cambridge, 1995

Articles

The following list does not include minor newspaper stories and magazine and journal reports mentioned in the text.

On Tunguska

V.A. Alekseev, 'New aspects of the Tunguska meteorite problem', *Planetary and Space Science*, vol. 46, 1998, pp. 169–77

I.S. Astapowitsch, 'On the fall of the great Siberian meteorite, June 30, 1908', *Popular Astronomy*, vol. 46, June–July 1938, pp. 310–17

William H. Beasley and Brian A. Tinsley, 'Tungus event was not caused by a black hole', *Nature*, vol. 250, 16 August 1974, pp. 555–56

V.A. Bronshten, 'Nature and destruction of the Tunguska cosmical body', *Planetary and Space Science*, vol. 48, 2000, pp. 855–70

John C. Brown and David W. Hughes, 'Tunguska's comet and non-thermal carbon-14 production in the atmosphere', *Nature*, vol. 268, 11 August 1977, pp. 512–14

Gavin J. Burns, 'The great Siberian meteor of 1908', *Popular Astronomy*, vol. 41, November 1933, pp. 477–79

Jack O. Burns, George Greenstein and Kenneth L. Verosub, 'The Tungus event as a small black hole: geophysical considerations', *Monthly Notices of the Royal Astronomical Society*, vol. 175, 1976, pp. 355–57

Andrew Chaikin, 'Target: Tunguska', *Sky & Telescope*, January 1984, pp. 18–21

William H. Christie, 'The great Siberian meteorite of 1908', *The Griffith Observer* (The Griffith Observatory, Los Angeles), vol. 6, April 1942, pp. 38–47

Christopher Chyba, 'Death from the sky', *Astronomy*, vol. 21, December 1993, pp. 38–43

Christopher F. Chyba, Paul J. Thomas and Kevin J. Zahnle, 'The 1908 Tunguska explosion: atmospheric disruption of a stony asteroid', *Nature*, vol. 361, 7 January 1993, pp. 40–44

Clyde Cowan, C.R. Alturi and W.F. Libby, 'Possible anti-matter content of the Tunguska meteor of 1908', *Nature*, vol. 206, 29 May 1965, pp. 861–65

Hall Crannell, 'Experiment to measure the antimatter content of the Tunguska meteor', *Nature*, vol. 248, 29 March 1974, pp. 396–97

J.G. Crowther, 'More about the great Siberian meteorite', *Scientific American*, May 1931, pp. 314–17

Walter Durranty, 'Kulik is returning from Siberia quest', *The New York Times*, 2 December 1928, p. 10 (see also 23 December 1928, sec. III, p. 3)

P. Farinella et al., '*Probable asteroidal origin of the Tunguska cosmic body*', *Astronomy and Astrophysics*, vol. 377, 2001, pp. 1081–97

Kirill P. Florenskiy, 'Did a comet collide with the earth in 1908?', *Sky & Telescope*, November 1963, pp. 268–69

Roy A. Gallant, 'Journey to Tunguska', *Sky & Telescope*, June 1994, pp. 39–43

Jack G. Hills and Patrick M. Goda, 'The fragmentation of small asteroids in the atmosphere', *The Astronomical Journal*, vol. 105, March 1993, pp. 1114–44

David W. Hughes, 'Tunguska revisited', *Nature*, vol. 259, 26 February 1976, pp. 626–27

A.A. Jackson and Michael P. Ryan, 'Was the Tunguska event due to a black hole?', *Nature*, vol. 245, 14 September 1973, pp. 88–89

L.A. Kulik, 'On the history of the bolide of June 30, 1908', trans. Lincoln La Paz, *Popular Astronomy*, vol. 43, 1935, pp. 499–504

—— 'The question of the meteorite of June 30, 1908, in Central Siberia', *Popular Astronomy*, vol. 45, 1937, pp. 559–62

Wolfgang Kundt, 'The 1908 Tunguska catastrophe: an alternative explanation', *Current Science*, vol. 81, 25 August 2001, pp. 399–407

—— 'The 1908 Tunguska catastrophe: a forming kimberlite?', The Tunguska 2001 International Conference Abstracts

Lincoln La Paz, 'Meteorite craters and the hypothesis of the existence of contraterrene meteorites', *Popular Astronomy*, vol. 49, 1941, pp. 99–102, 265–67

H.J. Melosh, 'Tunguska comes down to Earth', *Nature*, vol. 361, 7 January 1993, p. 14

263

George P. Merrill, 'The Siberian meteorite', *Science*, vol. 68, 11 May 1928, pp. 489–90

H.H. Ninniger, 'Contraterrene (?) meteorites', *Popular Astronomy*, vol. 49, 1941, pp. 99–102, 215–16

Chas P. Olivier, 'The Great Siberian Meteorite', *Scientific American*, July 1928, pp. 42–44

Ian Ridpath, 'Tunguska: the final answer', *New Scientist*, 11 August 1977, pp. 346–47

Z. Sekanina, 'The Tunguska event: no cometary signature in evidence', *The Astronomical Journal*, vol. 88, September 1983, pp. 1381–413

—— 'Evidence for asteroidal origin of the Tunguska object', *Planetary and Space Science*, vol. 46, 1998, pp. 191–204

Duncan Steel and Richard Ferguson, 'Auroral observations in the Antarctic at the time of the Tunguska event, June 30, 1908', *Australian Journal of Astronomy*, vol. 5, March 1993, pp. 1–10

Richard Stone, 'The last great impact on Earth', *Discover*, vol. 17, September 1996, pp. 60–71

V.V. Svetsov, 'Total ablation of the debris from the 1908 Tunguska explosion', *Nature*, vol. 383, 24 October 1996, pp. 697–99

—— 'Could the Tunguska debris survive the terminal flare?', *Planetary and Space Science*, vol. 46, 1998, pp. 261–68

Chris Trayner, 'Perplexities of the Tunguska meteorite', *The Observatory*, vol. 114, October 1994, pp. 227–31

R.P. Turco, 'An analysis of the physical, chemical, optical and historical impacts of the 1908 Tunguska meteor fall', *Icarus*, vol. 50, 1982, pp. 1–52

N.V. Vasilyev, 'The Tunguska meteorite problem today', *Planetary and Space Science*, vol. 46, 1998, pp. 129–50

'What a meteorite did to Siberia', *Literary Digest*, 16 March 1929, pp. 33–34

'The world's largest meteorite', *Literary Digest*, 30 June 1928, p. 19

F.J.W. Whipple, 'The great Siberian meteor and the waves, seismic and aerial, which it produced', *Quarterly Journal of the Royal Meteorological Society*, vol. 56, 1930, pp. 287–304

—— 'On phenomena related to the great Siberian meteor', *Quarterly Journal of the Royal Meteorological Society*, vol. 60, 1934, pp. 505–13

Kevin Zahnle, 'Leaving no stone unburned', *Nature*, vol. 383, 24 October 1996, pp. 674–75

Marek Zbik, 'Morphology of the outermost shells of the Tunguska black magnetite spherules', *Journal of Geophysical Research*, vol. 89, suppl., 15 February 1984, pp. B605–B611

—— 'Historical notes on the Tunguska catastrophe', *Bulletin of the Polish Academy of Sciences: Earth Sciences*, vol. 45, 1997, pp. 211–38

V.K. Zhuravlev, 'The geomagnetic effects of the Tunguska explosion', *RIAP Bulletin*, vol. 4, no. 1–2, January–June 1998

Others

John Abrahamson, A.V. Bychkov and V.L. Bychkov, 'Recently reported sightings of ball lightning: observations collected by correspondence and Russian and Ukrainian sightings', *Philosophical Transactions of the Royal Society, London*, Series A, vol. 360, 2002, pp. 11–35

Marcus Chown, 'Shadow worlds', *New Scientist*, 17 June 2000, pp. 36–39

Clifford J. Cunningham, 'Giuseppe Piazzi and the "missing planet"', *Sky & Telescope*, September 1992, pp. 274–75

Heber C. Curtis, 'The comet seeker-hoax', *Popular Astronomy*, vol. 46, no. 1, January 1938, pp. 71–75

Tom Gehrels, 'Collisions with comets and asteroids', *Scientific American*, March 1996, pp. 55–57

R.E. Gold, 'SHIELD: a comprehensive earth-protection architecture', *Advanced Space Research*, vol. 28, 2001, pp. 1149–158

Jeff Hecht, 'By Jupiter … what a performance', *New Scientist*, 23 July 1994, pp. 4–5

Samuel Herrick Jr, 'The "Phantom Bertha" mystery', *Popular Astronomy*, vol. 49, 1941, pp. 102–04

David H. Levy, Eugene M. Shoemaker and Carolyn S. Shoemaker, 'Comet Shoemaker-Levy 9 meets Jupiter', *Scientific American*, August 1995, pp. 69–75

Rosie Mestel, 'Night of the strangest comet', *New Scientist*, 9 July 1994, pp. 23–25

Andrei Ol'khovatov, 'Vitim bolide event', *Meteorite*, February 2004

'Shoemaker-Levy dazzles, bewilders', *Science*, vol. 265, 29 July 1994, pp. 601–02

D.J. Turner, 'Ball lightning and other meteorological phenomena', *Physics Reports*, vol. 293, 1998, pp. 1–60

Paul R. Weissman, 'The Oort cloud', *Scientific American Special Edition: New Light on the solar system*, 2003, pp. 92–95

Fred L. Whipple, 'Background of modern comet theory', *Nature*, vol. 263, 2 September 1976, pp. 16–20

Websites

Department of Physics, University of Bologna:
http://www-th.bo.infn.it/tunguska/

Southworth Planetarium, University of Southern Maine:
http://www.usm.maine.edu/~planet/tung.html

Professor Roy A. Gallant (University of Southern Maine) and others:
http://www.galisteo.com/tunguska/docs

Andrei Ol'khovatov, Moscow:
http://www.geocities.com/olkhov/tunguska.htm
http://olkhov.narod.ru/tunguska.htm

INDEX